RAINFORESTS OF THE WORLD

A Reference Handbook

RAINFORESTS
OF THE WORLD

A Reference Handbook

Kathlyn Gay

CONTEMPORARY WORLD ISSUES

ABC-CLIO

Santa Barbara, California
Denver, Colorado
Oxford, England

Library of Congress Cataloging-in-Publication Data

Gay, Kathlyn.
    Rainforests of the world : a reference handbook / Kathlyn Gay.
      p.   cm. — (Contemporary world issues)
    Includes bibliographical references and index.
    1. Rain forest ecology.   2. Rain forests.   3. Rain forest
conservation.   4. Deforestation—Environmental aspects—Tropics.
I. Title.   II. Series.
QH541.5.R27G39   1993   574.5'2642—dc20   93-39619

ISBN 0-87436-712-3 (alk. paper)

00 99 98 97 96 95 94 93   10 9 8 7 6 5 4 3 2 1 (hc)

ABC-CLIO, Inc.
130 Cremona Drive, P.O. Box 1911
Santa Barbara, California 93116-1911

This book is printed on acid-free paper ⊗ .
Manufactured in the United States of America.

Contents

2 Chronology, 29

3 Biographical Sketches, 41

4 Facts and Statistics, 67

5 Directory of Organizations, 111

6 Selected Print Resources, 137

7 Selected Nonprint Resources, 177

Preface

MANY PEOPLE WORLDWIDE ARE AWARE that tropical rainforests are being destroyed at a rapid rate. Throughout the year, 80 to 90 acres of tropical rainforest are destroyed every minute. An area of forest the size of New York state disappears each year. But not so well known is the major loss of rainforests in the temperate regions of the world, including western North America, New Zealand, Tasmania, Chile, Argentina, portions of Japan, northwest Europe, and the Black Sea coast of Turkey and Georgia.

Both temperate and tropical rainforests are vital parts of the earth's life support systems, supplying oxygen, regulating climate, generating precipitation, conserving soil, and providing ecosystems for multitudes of diverse species of plants and animals and homelands for thousands of indigenous people—those native to the land.

Over the past two to three decades, an increasing number of botanists, ecologists, climatologists, and other scientists have expressed their concerns about the diminishing forests of all types worldwide. Why are rainforests being obliterated? What are the benefits of rainforests to human populations around the world? What are the effects of rainforest destruction? What can be done to preserve rainforests?

These questions are basic to the research being conducted by those attempting to preserve rainforests and are just a few of the questions discussed in Chapter 1, which is an overview of rainforests worldwide. An effort to conserve forests, although not necessarily rainforests, began in the early 1800s in the United States, as the chronology in Chapter 2 indicates.

Biographical sketches of some leaders in rainforest preservation are included in Chapter 3, and Chapter 4 shows—with charts, graphs, and maps—the extent of the forest losses worldwide and the effects of destruction on people, plant, and animal life. Groups working to protect and conserve rainforests and to reforest devastated areas are the subject of Chapter 5.

Chapter 6 includes annotated bibliographic references on rainforests; rainforest animals, plants, and products; people who live in and depend on the rainforests for survival; the effects of disappearing rainforests; and efforts to save these valuable natural resources. Nonprint references are listed in Chapter 7, and a Glossary includes definitions of some of the terms associated with rainforests around the world.

Acknowledgments

A REFERENCE BOOK SUCH AS THIS is based on the prior work of many individuals—from academicians to volunteer workers—who have documented their efforts. For their dedication, a special thank-you. I would also like to thank the many people in a variety of organizations who provided information to include in this book and to express my appreciation for their commitment to programs and projects that help protect the remaining rainforests of the world.

I especially want to thank a cherished person, Doris A. Kimmel, for her help in preparing the annotated bibliography and her perseverance in the search for applicable titles to include in this book. Thank you, also, to Brian D. Byrn for his illustrations and to Douglas J. Gay for his adaptations of maps.

1

Rainforests of the World

An Overview

FOR DECADES, THE TERM "JUNGLE" has been linked with rainforests. Although "jungle" may describe the dense and lush growth of a forest, it is not a synonym for "rainforest." What, then, is a rainforest?

On a global scale, forest experts have classified dozens of different types of forests, and the categories for rainforests vary considerably because they may be classified by such factors as their location, soil type, climate, and whether they have remained in their "natural" state—undisturbed by human activities. Some rainforests cover lowland areas, others grow in mountainous regions, and still others spread across plains that are covered with floodwaters at certain times each year. However, the two main types of rainforests discussed throughout this book are tropical and temperate rainforests. As the terms suggest, the forests are located in either the tropical or temperate zones of the world.

A *tropical* rainforest may fit one of three basic types: a tropical dry forest, a tropical moist forest, or a true tropical rainforest. Forests in the latter category are closest to the equator and receive more rainfall than the other two categories of rainforests. In fact, calling a tropical dry forest a rainforest may seem a misnomer at times since rain may not fall for months at a time. Generally, a rainforest is defined by the amount of rainfall—from about 80 to 120 inches or more (sometimes much more) per year.

Tropical rainforests make up 6 percent of the earth's surface and are located in a 3.4 million-square-mile band on both sides of the equator, including parts of Central and South America, Africa, Asia, and the United States. Perhaps the most well known of all tropical forests is in Amazonia—the Amazon River Basin. Covering an area almost the size of the United States, the forest region stretches from the foothills of the Andes mountains to the Atlantic seacoast in Brazil. Other tropical rainforests include those in Papua New Guinea, the Congo Basin and Zaire in Africa, the islands of Madagascar, the province of Sarawak in Malaysia, northern Thailand, southeastern Mexico, and parts of Colombia and Ecuador. U.S. tropical rainforests include Wao Kele O Puna on the Big Island of Hawaii, the Caribbean National Forest in Puerto Rico, and rainforests on the three small islands of American Samoa.

Nearly all of the original rainforests in *temperate* zones were destroyed long ago, and because of varying classification methods and massive deforestation of temperate forests worldwide, there is limited information on the remaining stands of temperate rainforests. But some ecologists have attempted to assess the distribution and status of *coastal* temperate rainforests, classifying these forests as a subdivision of temperate rainforests.

U.S. Forest Service researchers Paul Alaback and Jim Weigand released published studies in the early 1990s that provided a provisional definition of coastal temperate rainforests. These forests are located in areas between 32 and 60 degrees latitude with at least 80 inches (2,000 meters) of annual precipitation (rain, snow, or fog). The researchers found three features "common to all coastal temperate rain forests: proximity to oceans, the presence of mountains, and as a result of the interaction of the two, high rainfall."[1]

Two conservation organizations, Ecotrust and Conservation International, supported the studies of coastal temperate rainforests and published a report in 1992 for scientific review. The report estimates that 80 to 100 million acres (30 to 40 million hectares) of coastal temperate rainforest once existed, but less than half of that remains, although no one is sure of the total. The forest areas that still exist are primarily on the western edges of continents and cover coastal regions in North and South America and also in New Zealand, Tasmania, portions of Japan, northwest Europe, Turkey, and Georgia (part of the former Soviet Union).[2]

In the southern hemisphere, the largest remaining area of coastal temperate rainforest is in Chile. Ecologists who have analyzed Chilean rainforests conclude that these forests "are the most species-rich in the world, due to a complex landscape which has produced diverse habitats and species composition. . . . One of the most impressive features . . . is the alerce cedars, possible relatives of the giant sequoias of California. The largest conifer in South America, these trees can reach 4,000 years of age, and form stands similar to the 'cathedral groves' of the Pacific Northwest."[3]

Of all coastal rainforests, those in North America are perhaps the most studied and inventoried. They account for about 40 to 50 percent of the world's remaining coastal temperate rainforests and stretch along a narrow, 15-mile wide strip of coastal land between the sea and the coastal mountain ranges from southern Oregon to southeastern Alaska, including Olympic National Park in Washington, coastal sections of British Columbia, the Carmanah Valley on Vancouver Island, and Alaska's Tongass National Forest. Some of these forests receive up to 200 inches of precipitation per year.

How Forests Affect Precipitation, Climate, Soil, and the Global Carbon Cycle

Wherever they are located, forests play a major role in generating rain by returning moisture to the atmosphere in a process known as evapotranspiration. That is, water from trees and plants growing in the forest evaporates and becomes part of the hydrologic, or water, cycle. About half the rainfall over Amazonia returns to the atmosphere through evapotranspiration. In this process, water also evaporates from oceans and other water bodies. Some of the water vapor then condenses to form clouds or changes into liquid or freezes to form crystals. In other words, it becomes dew, rain, snow, or other forms of precipitation.

Moisture in the atmosphere from evaporation affects regional and global climate. For example, clouds that form from moisture in the atmosphere cool forest regions. In addition, clouds over the tropics are carried by air masses circulating away from the equator to cooler regions of the earth, transporting solar heat with them to the temperate zone.

Forests in both temperate and tropical zones have elaborate root systems that hold soil in place, preventing erosion. The forest root system also absorbs rainfall, helping to regulate water runoff. Water is stored and slowly released throughout a year's time, replenishing groundwater supplies and maintaining the flow of water in rivers and streams.

The soil in tropical rainforests is poor, however. Unlike temperate forests in which trees draw nutrients from the soil, the tropical forest itself holds nutrients and exchanges them within or among vegetation, forest litter, and living organisms. Some tree leaves and branches obtain nutrients from rain as it falls or from plants and algae that grow on trees. When trees are cut, the few nutrients that remain in the soil soon wash out. Some of the soils also contain ironstone, a type of iron ore, and when this soil is exposed to rain it hardens like rock, preventing growth of most plant life.

Unlike tropical forests, the debris on temperate forest floors breaks down and nutrients are stored in the soil. As a result, after logging, fairly rapid regrowth may take place. However, soil erosion may prevent full recovery and the biological diversity of a logged area will not equal that of the original forest.

Forests are crucial in the cycling of the six basic elements—carbon, oxygen, nitrogen, hydrogen, phosphorous, and sulfur—that make up 95 percent of the world's living matter. Since there is a fixed supply of these elements, life on earth depends on their efficient cycling through the atmosphere and living systems. Carbon, one of the most important chemical elements, usually combines with other elements to form what seems to be an unlimited variety of compounds.

All living things are made up of carbon compounds, one of which is carbon dioxide (CO_2). Green plants absorb carbon dioxide from the atmosphere in a process known as photosynthesis. Chlorophyll, the green pigment in plants, captures sunlight and uses solar energy to combine CO_2 and water, producing a carbohydrate called glucose, an important source of energy for animal life, including humans. During photosynthesis, plants also release oxygen needed for survival. When people and other animals oxidize, or use up food, they exhale CO_2. Because of this continuous cycle, the supply of CO_2 and oxygen remains fairly stable.

Carbon dioxide is stored in various reservoirs such as the atmosphere and the oceans, in living plants and animals, and in fossilized forms such as oil, gas, and coal deposits. Driven by physical and biological forces, CO_2 circulates naturally among these holding places. But the amount of CO_2 held in the various reservoirs is changing because of human activities.

Forests are some of the major reservoirs for CO_2. Cutting down vast numbers of trees, whether in temperate or tropical forests, releases CO_2 into the atmosphere, adding billions of tons to the buildup of CO_2 from such activities as burning fossil fuels for energy, manufacturing, and transportation. As a result, some scientists theorize, CO_2 plus other gases have accumulated in the atmosphere, trapping heat that reflects from the earth, creating a condition known as the greenhouse effect, or an overall rise in the temperature of the planet. If global warming takes place, it could alter rainfall, wind, and heat patterns around the world.

Rainforests in North America

Often called "the emerald string," the coastal temperate rainforests in North America are dominated by evergreens such as western hemlocks, red cedar, white pine, Sitka spruce, and Douglas fir. Unlike the mass of vegetation that forms a closed canopy at the top of a tropical rainforest, the trees in North America's temperate rainforest tower like cathedral spires, narrowing at the top and allowing some sunlight to filter in.

Some of the trees, like the Douglas fir, may grow 80 to 200 feet (25 to 60 meters) tall, and provide enough timber to build a home. The largest Douglas firs are estimated to be from 400 to 1,000 years old and are among the world's oldest living things.

Broadleaf trees, which drop their leaves during winter, such as the bigleaf maple and red alder, can also be found in a temperate rainforest. Many of these trees are covered with mosses that do not feed off the trees but instead are nourished by moisture in the air.

A mass of living greenery carpets the understory, or floor, of a temperate rainforest. Flowers, ferns, and fungi abound. Some of the large mammals in the rainforest include elk, deer, mountain lion, black bear, beaver, and raccoon. Fish are plentiful in rainforest streams, and a variety of birds make their homes in the trees and shrubs.

Ancient, or Old-Growth, Stands

The Pacific Northwest rainforests include stands of ancient forests, sometimes called old-growth forests, although foresters and biologists define old-growth forests in a variety of ways. In the traditional view, an old-growth forest includes trees that have reached their peak in wood production. Using a broader definition, many people now say an old-growth forest is one in which there are large numbers of old conifers, including Douglas fir at least 300 years old and species of pines ranging from 150 to 3,000 years old. Ancient (old-growth) forests also contain snags (dead and rotting trees still standing) and many fallen logs, often called nurse logs, which provide homes and food for birds, mammals, and insects. As logs decay they provide nutrients for tree seedlings that may eventually spread their roots and straddle their nurse logs to reach the ground.

An old-growth forest is a complex interdependent system of living things. All parts of the system are interrelated and each organism depends on a part of the system for survival. If one part of the system dies or is damaged, the entire web of life is affected. For example, when lumber companies log in an old-growth forest, they frequently clear-cut, or cut hundreds of patches of about 40 to 60 acres (16 to 24 hectares) and clear off land for roads leading into the logging areas, a practice that has severely fragmented forest habitats. In 1992, the National Aeronautics and Space Administration (NASA) Goddard Space Flight Center released satellite photographs showing the extent of fragmentation in the temperate rainforests of Oregon and Washington. Compared to photos showing tropical deforestation in Amazonia, NASA pictures revealed thousands of clear-cut areas in the Pacific Northwest, which some scientists say has created far more damage than in Amazonia's tropical forest.

Effects of Clear-Cutting

Not only does clear-cutting in temperate forests cause soil erosion, but it also leaves the old growth on either side of the clear-cut vulnerable to adverse weather conditions. Wind penetrates the forest and dries out and damages root systems of trees. Deer and elk often feed in clear-cut areas, increasing their populations. When snow covers the open land, the animals enter the forest, where the ground is protected, to look for food. Because the animal population has increased, the forest areas are overgrazed and part of the ecosystem is destroyed. Fragmentation of the forest also splits up ecosystems and sometimes prevents some insect and other species from migrating because they have not been able to cross open areas.

Clear-cutting has been a major factor in controversies over the loss of wildlife habitats, particularly that of the northern spotted owl, an endangered species. To many environmentalists, the owl is an indicator species and has been likened to the canary that miners once carried underground to determine whether poisonous gases were present in a mine. If the canary died, the miners knew the mine was unsafe. In the case of spotted owls, which need old trees and space for survival, their diminishing numbers warn that the ecosystem is being destroyed.

Because no laws specifically safeguard endangered forests, those who want to preserve old-growth ecosystems must rely on laws that protect endangered species. Yet loggers fear their jobs will be lost if lumbering is banned in old-growth forests. Timber companies and loggers lash out at scientists and environmental groups working to save the forest habitats for spotted owls.

A "Medicinal Tree"

Another debate has centered on Pacific yew trees found in ancient forests. Once thought to be mere weeds or scrub growth, yew trees were usually burned with the slash—the undergrowth cut as loggers made their way to major stands of timber. But researchers discovered in 1989 that the bark of ancient Pacific yew trees produces a chemical known as taxol that has shown promise in treating ovarian and breast cancer and may also be effective in treating melanoma and other types of cancer. It takes three mature yew trees—about 150 years old and ten inches in diameter—to produce the bark needed to extract the drug for treatment of one patient. By the end of 1991, more than 900,000 pounds of yew bark had been collected on federal lands, yielding enough of the drug to treat 12,000 patients.

Many environmentalists and some government officials feared that because of high demand for its bark, the Pacific yew would be quickly harvested out of existence and that in the process, old-growth forests—the primary habitat for the yews—would be further endangered. But the U.S. Congress passed legislation, which President George Bush signed into law in 1992, requiring long-term conservation and efficient management of the yew in national forests. At about the same time, Weyerhaeuser, a major timber company, planted 2.6 million Pacific yew trees, an ornamental variety that produces the most taxol, in two nurseries, one south of Portland and the other near Olympia, Washington. And in 1993, the company planted 10 million yews in a Centralia, Washington, nursery.

The trees will be harvested for the anti-cancer drug in 1995, when they should be about two or three feet tall and one-half inch in diameter. The timber company expects to be able to provide enough trees to meet from 50 to 100 percent of the demand for taxol in the near future. It will be possible to meet that demand using the relatively young trees because of new technologies, including a process to synthesize taxol from every part of the tree— not only from bark, but also from twigs and needles—a process first developed in Italy and now replicated in the United States by the pharmaceutical company Bristol-Myers Squibb.

Tropical Rainforests of the World

Many people think of tropical rainforests as dark, intimidating places as they are so often depicted in movies. But people who explore and study rainforests frequently describe the "symphony of life," "plush vegetation," "magical scents," and "rich eternal rhythms" that are there.

Most tropical forests are in the warm, wet areas ten degrees on either side of the equator, where they flourish best. Their year-round temperature averages 80 degrees Fahrenheit (27 degrees Celsius).

Biodiversity in Tropical Rainforests

Because of the climate, tropical forests contain more diverse plants and animals than any other place on earth. At least 50 percent and perhaps up to 90 percent of all living species can be found in tropical forests.

An estimated 50 million animal species, including millions of insects, flourish in tropical rainforest areas. No one knows how many exist because scientists have only recently begun to classify and inventory species in various rainforest locations. An effort initiated in 1989 to categorize species in Costa Rican rainforests resulted in estimates of 500,000 different animal and plant species, among them an estimated 365,000 insect, spider, and tick species.

According to the National Academy of Sciences, insects in the rainforests are so abundant that a single tree in the rainforests of Panama may be home to more than 1,700 species. In the rainforest of Sarawak, Malaysia, biologists have identified 3,000 species of butterfly and moth.

Mammals that make the rainforest their home include various types of cats, deer, wild pigs, monkeys, sloths, marsupials, rats, and mice. Two-thirds of all known plants grow in tropical forests. "Of the 250,000 species of plants described by botanists, at least 30,000 are to be found in Amazonia alone," the World Rainforest Movement reported.[4]

Tropical Rainforest Habitats

Life in a tropical rainforest exists in layers of vegetation that form separate habitats, beginning at the forest floor and rising with the trees to well over 100 feet (30 meters). Emergent trees push above the canopy, a mass of tree tops intertwined with vines and colorful, flowering plants. The dense top tier of the forest also supports a diverse number of mammals, birds, and insects.

Another layer of growth called the understory begins about 50 to 60 feet above the ground, rising to a little more than 80 feet (25 meters). Since this tier is cooler and more humid than the canopy, fewer plants grow. Although the forest floor is usually bare except for fallen leaves, decaying plants, and some sprouting plants, this part of the forest is, in effect, a recycling center. Fungal enzymes help break down fallen leaves, and an army of ants, beetles, termites, worms, and many other organisms as well as foraging birds clean up the forest floor. At the same time, these creatures help renew the forest by planting seeds or providing nutrients for continued growth.

Products from Tropical Forests

Hundreds of items that people use on a daily basis in industrialized nations come from tropical rainforests. They include foodstuffs (fruits, vegetables, nuts, spices, teas, and ingredients for soft drinks), cosmetics, houseplants, fibers, rubber products, building materials, oils, perfumes, and pharmaceuticals. Thaumatin, a newly discovered compound derived from the katemfe bush, which grows in the West African rainforest, has been called the sweetest substance in the world—it is 100,000 times sweeter than table sugar, and it could be processed for commercial use as a sugar substitute. Another West African rainforest plant, the Calabar bean, is the source of compounds for some insecticides.

A number of rainforest plants and trees produce oil that can be used as fuel. Some nut trees produce a crude oil that can substitute for gasoline. Sap from Amazonian copaiba trees has the potential to become a substitute fuel for motor vehicles. Its properties are identical to diesel fuel and when poured directly into the fuel tank can be used to power a truck. (A chart of some other forest products is included in Chapter 4.)

Some of the most important rainforest products are medicinal drugs. Between 80 and 90 percent of indigenous populations in rainforest areas depend on plants for medicinal purposes. Nearly one-fourth of all pharmaceuticals that Americans use originally derived from tropical plants, which provided the chemicals for processing synthetic drugs. Quinine, for example, was developed from the bark of the chinchona tree and is used to treat malaria and pneumonia. Diosgenin, a primary ingredient in cortisone, is produced from rainforest plants that thrive in Belize, Guatemala, and Mexico. Curare from the Amazon is an ingredient for muscle relaxants. Drugs that stimulate the heart and respiratory system, anesthetics, tumor inhibitors, contraceptives, and anti-cancer medicines come from tropical rainforests. The Black bean (castanospermum australe) from the Queensland, Australia, rainforest is one of the world's great hopes for an AIDS cure.

Scientists have identified more than 2,000 tropical plants (some say the total could go as high as 3,000) that contain substances for treating various types of cancer. One example is the rosy periwinkle found in the rainforest of Madagascar, an island off the southeast coast of Africa. The periwinkle's tiny pink flower is the source of a compound for treatment of childhood leukemia and another compound for treatment of Hodgkin's disease. Madagascar's eastern forest, in fact, is said to be a "biotic treasure house," because 12,000 known plant species and 190,000 known animal species originated there. But at least half of the original plants and animals have been lost along with a vast portion of the forest—only about 10 percent of the original forest remains.

As forests disappear worldwide, more and more research efforts have been initiated to investigate the chemical properties of tropical plants. For example, the National Cancer Institute with headquarters in Maryland has sent ethnobotanists (scientists who study the interrelationships between plants and people) to various countries to find plants that native healers use for medicinal purposes. The institute and its researchers are concerned that not only the plants but also the healers will disappear before their curative secrets are revealed.

According to a 1992 report in *The Lancet,* a British medical journal, "The loss of chemical structures before they can be evaluated has been likened to the loss of the great libraries of Alexandria." Yet until the late 1980s and early 1990s, few pharmaceutical companies had conducted research on tropical plants. Now 223 companies worldwide, including 107 U.S. firms, are involved in such projects. One example is the U.S. company Merck, which invested $1 million in a research project in Costa Rica, home of 11,000 plant species. Ten percent of the funds were used for conservation of rainforests, and the remaining money was allocated to train local researchers. "Royalties from the sale of any product arising from the programme will be shared between Merck and Costa Rica," the magazine reported.[5]

Causes of Deforestation

In spite of the fact that forests and forest products provide benefits for many people, nearly 40 percent of the original tropical forests have been destroyed—cut or burned. At a snap of the fingers, an acre of tropical rainforest literally disappears. The United Nations Food and Agricultural Organization (FAO) estimated in late 1991 that the world's tropical forests were disappearing at a rate of 42 million acres (17 million hectares) per year, a rate 83 percent faster than that of forest loss between 1976 and 1980. Rainforests in temperate regions suffer great losses also, and some say temperate forests are disappearing at a more rapid rate than tropical rainforests. If this rate of destruction continues, most rainforests around the world will be gone within a century.

The causes of forest destruction vary by region but include excessive logging, road construction, urban development, slash-and-burn agricultural methods, mining, and economic policies that encourage exploitation of natural resources. In some regions all of these factors play a role in massive deforestation.

Reasons for U.S. Rainforest Destruction

The main cause of deforestation in the temperate zone is excessive logging, the construction of the many roads needed to transport timber, and timber slides—areas cleared to slide timber down mountain sides. Most virgin forests, which had existed for millions of years, have been cut down in industrialized countries.

America's rainforest once blanketed 70,000 square miles along the Pacific coast. But since the mid-1800s, when people began to push West to exploit the natural resources of the area, lumber companies have harvested most of the timber in the rainforest areas. Logging increased substantially after World War II to meet the demands of a growing population for housing, businesses, and other construction purposes.

According to the Wilderness Society, less than 30 percent of the 4.7 million acres (about 2 million hectares) of ancient forest that make up much of the U.S. coastal rainforest are protected in parks and wilderness areas. Federal laws allow logging on the remaining 3.3 million acres (about 1.3 million hectares) that belong to lumber companies or other private owners or are on federal lands managed by the U.S. Forest Service (USFS) and the Bureau of Land Management (BLM). Congress, which appropriates funds for management of national forests, determines the amount of timber to be harvested each year from these forests.

U.S. Government Policies

A share of the funds from the sale of timber harvested on federal land is turned over to counties where national forests are located. The payments compensate for property taxes that might have been paid by private owners. Other funds for maintaining and managing national forests come from such forest uses as mining, livestock grazing, and recreational activities, as well as receipts for the sale of salvage timber and compensation for building roads for private timber companies. Most of the funds collected from timber sales are retained by the Forest Service. But because timber and timber resources are frequently sold at below market rates, timber sales do not cover overall Forest Service expenses, a major portion of which is the cost of preparing for timber sales—about $125 million each year. Thus Congress appropriates millions each year to manage national forests.

Even though private landowners manage many forest resources at a profit and timber sales bring in well over a billion dollars in receipts annually, the timber program loses money each year—over $175 million in 1990 and $375 million in 1991. Overall, the USFS loses $1.5 billion annually, according to economist Randal O'Toole, who has been documenting low-cost timber sales and USFS operations since the 1980s. In short, taxpayers subsidize the sale of timber and also contribute to the destruction of temperate rainforests in the Pacific Northwest.

Influence of Timber Industries

Timber industries in the Pacific Northwest play a major role in continued rainforest loss through campaigns to convince their employees, federal officials, and the general public that jobs depend on increased logging. However, some state officials in Washington, Oregon, and California, leaders of environmental groups, and some business people point out that jobs in the timber industry have been declining steadily because of various factors:

1. Giant companies or financiers have taken over smaller firms and have sold off timber as fast as possible to pay off debts and provide funds to invest in other types of industries that show better profits than lumbering.
2. Many lumber mills have automated their operations, reducing the need for workers.
3. Rather than milling logs in U.S. mills, companies are shipping whole and split logs to Japan and other Asian countries and to Mexico, in effect exporting thousands of jobs along with timber.
4. The demand for timber from the Pacific Northwest has been decreasing since the early 1960s because lumber companies have been able to buy timber from other sources, some of which are woodlands in various parts of the nation that have been planted specifically for the timber produced.

Causes of U.S. Tropical Rainforest Loss

Vast amounts of U.S. rainforests in tropical regions have also disappeared because of logging and highway construction. In 1991, the USFS and the Federal Highways Administration announced plans to accept bids for rebuilding a road—originally constructed by the Civilian Conservation Corps during the 1930s depression—through Puerto Rico's Caribbean National Forest, which was once part of a rainforest that blanketed much of the West Indies. Known as El Yunque, the forest was designated a reserve by President Theodore Roosevelt and now covers only about 28,000 acres (11,331 hectares), one of the smallest forests under USFS jurisdiction.

A dozen environmental groups have been fighting the highway construction, pointing out that good roads already ring the forest and that a highway winding steeply through the mountainous area would go nowhere. Road construction would cause landslides and destroy large areas of habitat, further endangering species already at great risk. At least 225 species of trees grow in the forest, providing habitat for 1,200 species of insects and many mammals, reptiles, and birds. The endangered Arctic peregrine falcon winters in El Yunque, and the vanishing Puerto Rican parrot lives there permanently, although only 28 of these birds still exist in the wild.

Rainforests on the Virgin Islands, destroyed years ago for sugar and cotton plantations and cattle ranches, have partially grown back but now are threatened by home construction and resort development. Tourism and resort development also threaten rainforests on the islands of American Samoa, although the U.S. Congress designated two of the islands to be part of a new national park in order to protect forest and reef areas.

On the Big Island of Hawaii, a geothermal project has threatened Wao Kele O Puna, the last lowland tropical rainforest in the United States. The rainforest grows on the flanks of the highly active Kilauea volcano where an electrical company wants to generate steam power from geothermal wells drilled into the volcanic rock. A road has been constructed in the area and land has been cleared for drilling. Plans are for several hundred shafts to be sunk more than a mile beneath the surface.

Drilling devastates more than plant and animal species. It also violates native Hawaiian religious beliefs, since it is a desecration of the Goddess Pele who is embodied in the volcano. Hawaii Electric Industries was able to obtain the land area as a result of a questionable land exchange that contained no provisions for an environmental impact statement. Noxious hydrogen sulfide, which would be released from the proposed geothermal wells, is highly poisonous and could destroy plant life and pose serious health risks for human beings. Brine from the wells would be released on the surface and, after killing the surface vegetation, would percolate down to contaminate fresh water, the island's lifeblood.

Federal funding was appropriated for the geothermal project, and state and federal officials continually reassured residents living near the drilling area that all operations were safe. But in June 1992, two explosion from drilling operations occurred, leaving several dozen animals dead, injuring two workers, and causing breathing difficulties and other health problems for nearby residents. Because of the explosions, government officials halted drilling operations until safety could be assured, and a federal judge froze federal funding for the project until completion of an environmental impact statement.

Reasons for Deforestation in Developing Countries

In the nonindustrialized nations of the tropics, wood is the primary energy source for millions of poor and landless rural people. Thus many peasants cut down trees to provide fuelwood needed for cooking and heating. Fuelwood harvesting depletes 5 million acres (about 2 million hectares) of tropical rainforest every year.

However, as in temperate zones, commercial logging has a much more devastating impact. Timber exports are major sources of income for many nations in tropical regions, and tropical woods have been shipped primarily from Africa and southeast Asia to Japan, the largest consumer of tropical timber (the United States is second). Japan uses much of its imported tropical timber to make such products as chopsticks, toothpicks, and furniture. The country also uses wood chips to make cardboard packaging for electronic equipment. Originally those chips came from the United States, but since large quantities of U.S. waste chips are no longer available, Japan has set up its own operations in Papua New Guinea, clearing hundreds of acres of rainforest on the north coast.

The timber industry has caused extensive damage and destruction of rainforests around the world not only because trees are cut but also because trees left standing are injured. For every 26 trees cut in Sarawak, Malaysia, for example, 33 trees are damaged and 70 percent eventually die.

Destructive Social and Economic Policies

In its Emergency Call to Action to save tropical forests, the World Rainforest Movement declared that social and economic policies of so-called Third World, or nonindustrialized, nations are to blame for deforestation. In many of these countries, the major cause of deforestation is a government development strategy designed to produce crops for export, earning funds to pay off huge debts to industrialized nations and to buy manufactured goods.

Vast areas of forestland in Malaysia, for example, have been cleared and cultivated to produce rubber, palm oil, and cocoa. According to Malaysian officials, the deforestation rate has risen dramatically. During the period from 1981 to 1985, 3.7 million acres (about 1.5 million hectares) of rainforest were demolished each year. In 1989, 9.51 million acres (3.5 million hectares) of forestland were stripped of trees.

Most of the fertile land areas in Amazonia and the Philippines have been turned into huge plantations, many owned by multinational companies, to grow such export crops as bananas, pineapples, sugar cane, and rice. As a result, large numbers of peasants, or poor farmers who own no land have little or no access to fertile soil. They are forced to cut or burn forests and clear the land to grow their own food. But the soil degrades within a few years and little grows. Peasants must then move on to other land, repeating the slash-and-burn practices again and again just to survive.

There are many views on how the destruction of the Amazon rainforest began. But in the opinion of Susanna Hecht, an expert on the Amazon, "The causes of environmental degradation in the Amazon can be traced to a philosophy and strategy for regional development formulated by the Brazilian military,"[6] which gained great power and influence beginning about the 1930s and expanded rapidly after World War II. The Brazilian military pressed for the transfer of public lands to private landholders—primarily those with wealth and power—and for the development of roads through rainforest areas, displacing indigenous people and killing thousands through bloody conflicts.

A military junta took over the Brazilian government in 1964 and began to implement plans to settle Amazonia, which made up half the nation's land area but was home for only 3.5 percent of Brazil's population of 70 million. By encouraging people in densely populated areas of Brazil to colonize Amazonia, the military government, with the support of churches and other organizations, hoped to provide land and "vibrant opportunities" for those who migrated and settled in the Amazon. In the words of Hecht, church leaders "exhorted their charges to seek the promised land" and the Brazilian government "offered tangible inducements: transportation to the Amazon and a 240-acre plot for each settler, with sure title, guaranteed credit for the planting of rice, corn and beans; a six-month household subsidy to tide the family over the initial difficult months; and food subsidies as insurance against disasters. Colonists were also promised housing, schools, medicine, transportation and technical assistance."[7] But many colonists failed in farming efforts because poor soil produced few crops or pests destroyed crops.

Beginning in the late 1960s, the Brazilian government implemented a plan to build roads and encourage commercial investment in the Amazon, offering financial incentives such as a moratorium on corporate income taxes for a dozen years, subsidized loans, and reduced import duties for manufacturing or agricultural equipment needed to develop Amazonian land.

Cattle Ranching

The Brazilian government encouraged cattle ranching as a means of developing. But little beef has been exported from Brazil because cattle are plagued with foot and mouth disease. Furthermore, only 3 out of 100 ranches make any profit from ranching, even with government subsidies. Yet cattle ranching has grown rapidly. Why? According to the World Rainforest Movement:

> Under Brazilian law, anyone who clears an area of forest can lay claim to the land. Cattle permit large amounts of land—and the mineral rights below it—to be claimed with minimal labour. It is thus no coincidence that those areas where clearance is most vigorous are frequently closest to gold strikes. It is this quest for quick (and easy) profits—rather than a hunger for beef—that provides the key to much of the deforestation in Brazil. Indeed, at least half of the largest ranches in Brazilian Amazonia have never even sent a cow to market.[8]

In Central America, however, cattle ranching is a primary cause of deforestation. Ranchers clear rainforest land to provide pasture for their cattle, but as is the case with farming schemes, the soil is soon depleted and pasture grass does not grow, so more rainforest land is cleared. As Jeremy Rifkin, an environmental activist, noted in his book *Beyond Beef:* "Since 1960 more than 25 percent of the forests of Central America have been cleared to create pastureland for grazing cattle." Most of the beef produced is exported to North America for consumption as hamburger in fast food restaurants.[9]

Wealthy landowners and international corporations and banks are the main beneficiaries of cattle raising, taking over the productive land and leaving marginal land for poor rural families. In Costa Rica, for example, 2,000 ranching families own more than 50 percent of the productive land. Wealthy landowners control 70 percent of agricultural land in Colombia, using much of it for cattle grazing.

Oil Extraction

Drilling for oil is endangering one of the most biologically diverse rainforests in the world—the Oriente in remote regions of Ecuador. Although a large portion of the Ecuadoran rainforest, including Yasuní National Park, has been set aside as a reserve, oil companies have ravaged about one-tenth of the forest since the 1970s, drilling hundreds of wells and constructing a network of roads and pipelines. A 280-mile pipeline carries heavy grade crude oil to ports where it is pumped into tankers and transported to refineries around the world—at least half of the exports go to the United States.

Numerous international environmental groups and some indigenous groups have tried to stop further oil exploration, but not much progress has been made. The Ecuadoran government, which receives revenue from the oil extraction, has allowed continued exploitation of rainforest areas in order to pay off some of its huge foreign debt.

Oil exploration not only destroys vast amounts of rainforest. It also contaminates soil and drinking water supplies as extremely toxic wastes from oil production are dumped into open pits, creeks, and streams, threatening the health and lives of indigenous people. One investigative reporter visiting Lago Agrio, a so-called boomtown near Amazon oil country, described Lago as

"a full-blown slum in the midst of what was only twenty years ago virgin jungle. There are open sewers, rooting pigs, piles of garbage on the unpaved rutted streets . . . dilapidated housing, sooty tire-recapping stands, used auto-part stores . . . hookers displaying their thighs in open doorways, two-dollar-a-day oil workers knocking back their second six-pack by noon, and a prevailing mood of restlessness, tension, and frustration. . . ."[10]

Dam Building, Mining, and Industrialization

Hundreds of thousands of square miles of rainforest have been lost due to dam building and flooding for reservoirs that provide hydroelectric power. Governments in countries such as India, Indonesia, and Brazil have encouraged dam building in order to support new industries, including mining activities and foundries, and to provide irrigation. But the dams and industrialization offer few benefits to the rural poor or tribal groups. Because of dam projects, people have been displaced, which in turn destroys cultures and condemns many to life in urban slums.

The Impacts of Deforestation on Indigenous People

In tropical regions, rainforests are home for an estimated 50 million indigenous people—those native to the land—who make up more than 1,000 distinct groups. Seven hundred tribes live in Papua New Guinea alone, and about 200 tribes make Africa's Congo Basin their home. Other tribal groups include the Penan of Malaysia, the Efe of Zaire, the Lua of northern Thailand, the Hanunoo in the Philippines, the Kayapo of Brazil, the Sanema of Venezuela, and the Lacandon Maya in Mexico.

Each indigenous group has its own way of life, but all rely on the forest for some if not all aspects of their survival. Consider the Sanema of Venezuela, a semi-nomadic people who are the most isolated of rainforest groups in South America. The forest is their life—it provides them with all their physical needs such as food and shelter as well as spiritual sustenance. *The Gaia Atlas of First Peoples* describes the Sanema agricultural system as "undemanding" and sustainable:

> The Sanema plant crops, such as cassava, bananas, cocoyams, and plantains, that grow easily in the nitrogen-poor soils. And they move often to allow the exhausted soil to recover. The low-protein garden produce is usually complemented by meat, obtained by hunting with bows and arrows within an 8-km (5-mile) radius of their main village. Integral to the Sanema economy are long periods when they abandon their gardens and trek in the forest, setting up temporary shelters as far as 25km (15 miles) from the village, in order to hunt and collect forest foods, which minimizes the burden on any one area of land.[11]

Because of deforestation, many indigenous cultures and lives of tribal members have been threatened. Tribal groups such as the Penan of Sarawak, Malaysia, are hunters and gatherers, sustained by the rainforest plants and animals. However, only a few hundred are able to live in traditional ways as Malaysian forests are logged extensively. Since 1989, the Penans have tried to prevent further logging by barricading roads into the tropical forests. But government officials have arrested many of the tribal people and some Penan leaders have had to flee the country. Most of the tribe, numbering about 10,000, have been forced to live in villages and cities because their homelands have been devastated.

In Brazil, the livelihood and lives of indigenous groups were threatened when the national government planned to flood up to 97,000 square miles (250,000 square kilometers) of land for dam projects to be financed by loans from the World Bank. Out of 168 planned dams, 68 were on territories claimed by indigenous groups. But environmentalists in Brazil and other countries encouraged leaders of the Kayapo tribe to launch a campaign to stop the dam-building projects. Kayapo leaders traveled to World Bank headquarters where they asked officials not to fund the dam-building projects. They also met with federal legislators in the United States and Europe. In addition, the Kayapo held an international meeting at one of the dam sites, which resulted in worldwide publicity and pressure on Brazil to withdraw its loan proposal, halting the dam project.

Indigenous people in many countries have died because of contact with outsiders—usually whites of northern European extraction—who have brought contagious diseases, ranging from measles to influenza, and sexually transmitted diseases. More than 30 percent of the Yanomami population in Venezuela, for example, has died from whooping cough and measles. Air, soil, and water pollutants from industries established by Asian, European, and American companies in tropical forest areas have also posed serious health hazards for indigenous groups.

Wherever the culture and lives of indigenous people are extinguished, there is a loss of thousands of years of accumulated knowledge of natural ecosystems. This in turn affects industrialized societies. People in so-called advanced nations can learn much from the world's native peoples about how to use the land and harvest products without destroying the environment. For example, the Kayapo of Brazil's rainforests plant gardens in complementary groupings, a practice now popular in industrialized nations. In some instances, certain plants thrive when growing beside each other; in others, some plants release poisonous substances that discourage various insects that threaten a nearby plant. The Kayapo also fertilize their garden plots with rotting plants and soil rich in nutrients from ant and termite nests. In addition, the Kayapo gather and make use of hundreds of plants for medicinal purposes.

Effects of Deforestation on Natural Resources

The highly probable loss of valuable medicinals from rainforests is a major consequence of clearing off rainforest land. But not only diverse plant life is being threatened. Many mammal, bird, insect, and other species are becoming extinct, too.

Species Loss

According to the World Resources Institute's 1992 *Environmental Almanac*: "Because of the leap in deforestation rates since 1950, species loss has increased significantly since that time. If one considers plants and anthropoids along with birds and mammals, estimates indicate a potential loss of between 5 and 15 percent of the world's species between 1990 and 2020, or 50 to 150 species per day."[12]

Many ecologists say that such a species loss has not occurred since the dinosaurs became extinct 65 million years ago. Biologists Paul and Anne Ehrlich have likened the loss of individual species, from bacteria to mammals, to the continued loss of rivets that hold an airplane together. While the short-term effects may not be noticeable, the long-term effects could be disastrous.

When a species becomes extinct, genetic resources disappear too, which "diminishes humanity," in the words of world-renowned biologist Edward O. Wilson. In an article for *Scientific American*, Wilson explained:

> Every microorganism, animal and plant contains on the order of from one million to 10 billion bits of information in its genetic code, hammered into existence by an astronomical number of mutations and episodes of natural selection over the course of thousands or millions of years of evolution. Biologists may eventually come to read the entire genetic codes of some individual strains of a few of the vanishing species, but I doubt that they can hope to measure, let alone replace, the natural species and the great array of genetic strains composing them. . . . Species diversity—the world's available gene pool—is one of our

planet's most important and irreplaceable resources. No artificially selected genetic strain has, to my knowledge, ever outcompeted wild variants of the same species in the natural environment.[13]

Impact on Agriculture

The loss of species can have an adverse impact on the world's agriculture. Many food crops grown in the United States and other industrialized nations have been improved through cross-breeding with tropical varieties of plants, which has resulted in crops that are more resistant to disease.

Insects from the tropics are being used in biological pest control programs. That is, beneficial insects act as predators or parasites, destroying insect pests—those that are harmful to crops. Many harmful insects have become resistant to pesticides manufactured from artificial chemicals, so in some cases the use of beneficial insects has proven to be the most effective way to reduce or eliminate pests. Natural pest controls also help cut down on the use of chemical pesticides that pose health hazards to farm workers who apply them and to those who use water supplies contaminated with pesticides or eat foods with pesticide residues. As forests are destroyed, many beneficial insects disappear as well.

Deforestation is responsible for the destruction of forest products such as fruits, nuts, and medicinal plants. In addition, when trees are cut or burned, the forests' sponge effect is lost, and flooding occurs. Extensive logging caused flooding of massive land areas in Thailand in 1988 and brought much death and destruction. The Ganges Plain, the most densely populated region in the world, has suffered the most severe flooding because forests in India and Nepal have been cleared for agriculture.

Flooding, Erosion, Desertification, and Climate Changes

Logging vast areas leads to denuded land, which erodes and increases the sediment in rivers, in turn destroying habitats for fish and other aquatic life. Silt from runoff of topsoil slows water flow and reduces the efficiency of hydroelectric dams. Sediment also adversely affects water quality. When land is stripped of trees, desertification results—that is, cleared land becomes desert.

Changes in rainfall could be another serious consequence of deforestation. In Panama, for example, the rainfall pattern over undisturbed forest areas has remained fairly constant. But in areas where forests have been cut, the rainfall has decreased steadily over the past 50 years. Tropical deforestation could result in decreased rainfall and droughts over much of the food-growing areas of the United States and the former Soviet Union.

Deforestation contributes greatly to the problem of global warming, many scientists theorize. Large-scale burning of forests, as occurs in Amazonia, and extensive logging in temperate rainforests release large amounts of carbon dioxide that interact with other chemicals in the atmosphere, creating a buildup of so-called greenhouse gases that could increase the overall temperature of the planet.

Global warming could alter rainfall, wind, and heat patterns around the world. If the earth warms, polar ice caps could melt, resulting in sea level rise, which would flood many coastal areas, including some along U.S. coasts. Coastal flooding would endanger many people since some of these areas are heavily populated. The agricultural belts of the world could be altered, severely disrupting the economies of food-exporting nations like the United States. Given the predicted rate of temperature change, many species would be unable to adapt. They might have to migrate to more suitable environments, but migration could be blocked by urban development and other human activities that destroy habitats.

Efforts To Preserve Rainforests

Because of the many threats to rainforests and the predicted consequences of forest loss, hundreds of nongovernmental organizations (NGOs), including international environmental and human rights groups, and some government agencies around the world have been working to preserve both temperate and tropical rainforests. Indigenous groups in some countries also have joined together to demand changes in national policies that have destructive effects on rainforests. Some of these indigenous organizations have gained title to forestlands and have been managing them, as they have done for many thousands of years.

Numerous environmental groups worldwide call for boycotts of rainforest timber in order to slow down forest devastation. In Germany, for example, a coalition of students from 80 cities campaigned in 1991 to stop the disastrous effects of tropical timber consumption. They visited a variety of businesses asking merchants not to sell products made from tropical hardwood. Although some merchants were indignant and ordered the students out of their stores, the campaign had some effect. Some German importers announced they would stop importing Malaysian timber because of the impact of logging on the Penan. More than 450 local government councils in Germany also have responded to calls for boycotts and have banned the use of tropical timber in public works. Similar bans are in force in 90 percent of the local governments in the Netherlands, and in some U.S. city and state government agencies.

Some groups support debt-for-nature swaps, a program in which environmental organizations help pay off a portion of a nation's debt in exchange for rainforest land, setting it aside as a reserve. Other groups promote the sale of products from rainforests, which helps prevent the destruction of trees. For example, the environmental group Conservation International has initiated a variety of community development projects such as one in Ecuador to harvest and market tagua, an ivorylike palm nut that is used to make buttons, jewelry, watches, carvings, and other products. The tagua nuts are sold directly to factories, providing income for local workers and helping to create higher demand and prices for the nuts.

In some countries, local people have been collecting or raising rainforest animals and insects as a way to earn a livelihood in lieu of clearing the land for crop production. As a result, they also preserve forestlands. Butterfly farms are among some of the successful operations that prevent deforestation in Papua New Guinea. The butterflies under "cultivation" are the endangered homerus swallowtail known as birdwings. U.S. researchers and conservationists support the butterfly farms in order to prevent illegal poaching by collectors and to preserve the insects' habitat.

Local farmers in Panama and Costa Rica have been raising another type of rainforest "livestock"—the iguana, a green lizard known as the "chicken of the trees" because it tastes somewhat like chicken. Iguana meat is popular among many Latin Americans, but the lizards are disappearing because their habitats—tall trees—are being cut down. By developing iguana breeding areas, people can not only earn income from the meat sold but also help keep tropical trees standing.

Other promising conservation efforts are under way in Belize. This Central American country has been able to preserve more than 80 percent of its rainforests by establishing a variety of industries that can exist in harmony with the environment. One of the most important is ecotourism, which promotes tours and accommodations that respect nature and the environment. Another is the Belizean Rainforest Products, Ltd., which produces and markets potpourri and colognes made from rainforest ingredients such as individually handpicked allspice leaves and cultivated flowers.

Costa Rica also has been increasingly active in the preservation of rainforests and biodiversity. As in Belize, ecotourism is one effort that has been instrumental in saving rainforests. Local people have developed tourist businesses that have brought in thousands of visitors each year, helping to provide income without devastating forestland. Hundreds of scientists and ecologists also visit and use Costa Rica's rainforest reserves as laboratories to study plant and animal species and their habitats. In 1989, Costa Rica created the National Biodiversity Institute (INBio), where scientists can house specimens and catalog rare plants and animals found in the rainforests. At the same time, INBio hires local people to collect and help with the biological inventory.

In 1990, Brazil's President José Sarney signed laws providing for extractive reserves to protect more than 5 million acres (2 million hectares) of forest areas to be managed by rubber tappers, nut gatherers, and others whose livelihood depends on the rainforest harvest. After Fernando Collor de Mello was elected to office that same year, he appointed an internationally recognized environmentalist, José Lutzenberger, as Secretary of Environment, a newly created post. Lutzenberger had helped call attention to the World Bank's role in providing loans for projects, such as road building, that directly resulted in rainforest destruction.

Considerable business, political, and military pressure forced Lutzenberger's resignation two years later, but his influence may have been partially responsible for the Brazilian government's response to activists groups worldwide who long had called for preservation of rainforest homelands for indigenous peoples. In November 1991, the Brazilian government issued a directive setting aside a 36,000-square-mile (93,200-square-kilometer) reserve for the Yanomami and a 19,000-square-mile (about 30,600-square-kilometer) reserve for the Kayapo. Although segments of the military and some politicians threatened to block efforts to set up the reserves, the Brazilian Congress appropriated $2.8 million to mark the territories.

93-1309

Governments in other countries have also established preservation policies. The nation of Guyana on the eastern coast of South America has set aside 889,200 acres (360,000 hectares) of rainforest for conservation and research. New Zealand has banned the export of its forest products unless they come from a forest that is managed sustainably. In Vietnam and India, governments have launched forest conservation programs.

Deforestation and other forestry issues have gained attention on the international level. During the early 1980s, the United Nations Development Program and the UN Food and Agriculture Organization made an assessment of tropical forests and along with the World Bank and the World Resources Institute developed a Tropical Forestry Action Plan (TFAP).

A framework for actions, TFAP was implemented in 1985 and has received strong support from at least 80 national governments representing 90 percent of the tropical forest area. But many NGOs have criticized the plan, arguing that it does not take into account the effect of deforestation on climate change and biodiversity and the need for involvement of local communities.

In 1990, the UN General Assembly established the framework for the United Nations Conference on Environment and Development (UNCED), and for the next two years organizers planned the agenda for a global environmental meeting, declaring that talks about preservation of rainforests would be part of that agenda. As Maurice F. Strong, UN Secretary-General, noted: "the question of development and conservation of forests has moved from the professional and technical level to the international political agenda" because of "awareness of the critical role that forests play in maintaining the health and stability of the Earth's environment." Speaking to a World Forestry Conference in 1991, he pointed out:

> The harvesting of forests for immediate economic gain, if it is
> done in an unsustainable fashion, results in net destruction of
> the resource itself and exacts heavy environmental costs as
> well as economic costs which often greatly exceed the immediate
> economic benefits. Of course, sustainable utilization of forests is a
> valid and often necessary part of development, a point which is
> often made by countries with large virgin forest areas. Let us be
> clear—and realistic—until nations and people that depend on
> forests for their economic well-being can see that their economic
> interests are served by sustainable development of forests, the
> destruction of forests will continue.[14]

When UNCED, dubbed the Earth Summit, convened in Rio de Janeiro in June 1992, representatives of more than 150 national governments attended. The gathering marked the 20th anniversary of the first global environmental conference in 1972—the UN Conference on the Human Environment held in Stockholm, Sweden, which some observers have called the beginning of the modern environmental era.

Some participants in the Earth Summit hoped to reach global agreement on numerous environmental issues, among them agreement on how to counteract deforestation worldwide. Heads of state signed five documents, the most massive being a 40-chapter, 800-page volume called "Agenda 21," that includes detailed plans and projected annual costs for sustainable development programs. One chapter in the volume, "Combating Deforestation," is a discussion of various means to prevent rainforest destruction.

Other documents signed were a treaty on climate change and another on biodiversity; a statement of principles (the "Rio Declaration") on general environmental protection; and a statement of forest principles. Various international and national governmental organizations as well as environmental groups had urged that a treaty on protection of forests worldwide be negotiated during the Earth Summit, but the signed agreement on forest principles is a nonbinding declaration. It does, however, reflect global consensus on some forest issues and provides guidelines for conserving and managing forests in a sustainable way.

Notes

1. *Coastal Temperate Rain Forests: Ecological Characteristics, Status and Distribution Worldwide* (Portland, OR: Ecotrust and Washington, DC: Conservation International, June 1992), 3.

2. Ibid.

3. Ibid, pp. 22–23.

4. World Rainforest Movement, *Rainforest Destruction: Causes, Effects & False Solutions* (Penang, Malaysia: World Rainforest Movement, 1992), 16.

5. Linda Fellows, "What Are the Forests Worth?" *The Lancet* (May 30, 1992), 1332.

6. Susanna Hecht and Alexander Cockburn, *The Fate of the Forest: Developers, Destroyers and Defenders of the Amazon* (New York: HarperCollins, 1990), Chapter Six.

7. Ibid, 124–125.

8. World Rainforest Movement, *Rainforest Destruction: Causes, Effects & False Solutions* (Penang, Malaysia: World Rainforest Movement, 1992), 46.

9. Jeremy Rifkin, *Beyond Beef* (New York: Dutton/Penguin Books, 1992), 192.

10. Marc Cooper, "Rain-Forest Crude," *Mother Jones* (March/April 1992), 44.

11. Julian Burger, *The Gaia Atlas of First Peoples* (New York: Anchor Books Doubleday, 1990), 26.

12. World Resources Institute, *The 1992 Information Please Environmental Almanac* (Boston: Houghton Mifflin, 1992), 280–281.

13. Edward O. Wilson, "Threats to Biodiversity," *Scientific American* (September 1989), 114.

14. Maurice F. Strong remarks at United Nations Conference on Environment and Development at 10th World Forestry Congress, Paris, France, September 17, 1991, released via EcoNet, electronic UNCED conference, December 18, 1991.

2

Chronology

ALTHOUGH THE PROCESS OF DEFORESTATION worldwide has acceler-
ated dramatically since about the 1950s, the world's forests have
been exploited for centuries as people have cut trees for fuelwood
and have cleared land for agriculture, building, and road construction. During the colonial period in the United States, for
example, about 100 million of approximately 850 million acres of
original forested land were cut or burned, primarily for agricultural purposes. Between 1850 and 1900, an average of 13.5 square
miles of U.S. forest were cleared every day to make room for
farms. Forests also were cut for timber supplies used to meet
construction and fuel needs, leaving about 580 million acres of
forest in the United States by 1900.

As U.S. forests rapidly disappeared, some conservationists
began to speak out and write about the dangers of forest loss. This
chronology begins with the U.S. conservation movement, which in
recent decades has had a major impact on efforts to call attention
to tropical rainforest destruction worldwide and since the 1980s
has helped spotlight the degradation of temperate rainforests.

The U.S. conservation movement is said to have been inspired by George Perkins Marsh and his book *Man and Nature,*
published in 1864. In his book, Marsh pointed out the dangers of
deforestation and human interference "with the spontaneous arrangements of the organic or the inorganic world." Marsh warned
that the destructive effects of excessive logging in Europe could be
replicated in the United States as vast fertile regions in the East
and Midwest were being destroyed at an unprecedented rate.

In general, Americans, like people in other parts of the world, were not interested in preservation of forests and other natural resources. During the 1800s, Europeans and Americans were experiencing the industrial revolution and were more concerned about expanding industries, transportation, agriculture, ranching, and other human activities that contributed to deforestation. So conflicts developed between those who wanted to conquer nature and exploit natural resources such as forests and those who wanted to conserve or use resources wisely.

This chronology reflects the ongoing conflict over use of natural resources, particularly the timber from all types of forests, including temperate and tropical rainforests. It traces the expanding U.S. conservation efforts that eventually began to help raise awareness of rainforest destruction. The most recent years of the chronology reflect the growing understanding worldwide of the need to manage rainforests in a sustainable way, to protect biodiversity and forest products so that they will continue to provide basic needs for people around the world and particularly for those who make the rainforests their home.

1875   Conservationists in the United States establish The American Forestry Association to promote the value of trees and forests. The organization will continue to expand and eventually establish the Global ReLeaf program to plant trees in deforested areas of the world.

1882   A rider is added to the U.S. General Revision Act, authorizing the President of the United States to set up "forest reserves on the Public Domain."

John Muir, an early U.S. conservationist, establishes the Sierra Club to protect the natural habitats of the Sierra Nevada range in the western United States. The club will expand over the years to become one of the major U.S. conservation organizations promoting protection of wilderness areas, including rainforests.

1895   Gifford Pinchot, America's leading advocate of forest and other natural resource conservation, becomes head of the Division of Forestry under the Department of Interior.

1900   German-born Frederick Weyerhaeuser travels from the eastern United States to Tacoma, Washington, where he establishes a timber company that today owns 5.6 million acres in the Pacific Northwest where most remaining temperate rainforests exist.

1902   An alliance of forest product manufacturers establishes the National Lumber Manufacturers Association, which in 1965 will change its name to the National Forest Products Association, to lobby the U.S. Congress on behalf of the forest products industry and to present the manufacturers' views on how forests should be managed, views that will sometimes be in conflict with forest conservation ideas.

1905   President Theodore Roosevelt names Gifford Pinchot chief forester of the newly formed U.S. Forest Service (USFS), which had been transferred from the Department of the Interior and placed under the direction of the Department of Agriculture. Pinchot promises to "advocate nothing in the way of forestry that will not pay"—that is, to manage the forests so they will be self-supporting.

1916   The National Park Service is established in the United States. Now an agency under the Department of the Interior, the National Park Service manages national parks, some of which are temperate rainforests, and also oversees reserves, historic sites, parkways, trails, and similar federally owned recreational areas.

1930   The U.S. Congress passes and President Herbert Hoover signs the Knutson-Vandenberg Act, which gives the U.S. Forest Service the authority to require anyone buying timber from national forests to deposit with the federal treasury enough funds to cover the costs of reforestation (and in a later amendment, the cost of restoration of wildlife habitats) of the cut area. The act is designed to restore forest areas and in later years is frequently cited by groups working to protect ancient forests that are part of coastal temperate rainforests.

1936   U.S. President Franklin Roosevelt convenes the first North American Wildlife Conference to create a private organization designed to restore and preserve the continent's disappearing wildlife resources. The organization will be named the General Wildlife Federation, which will be changed in 1938 to the National Wildlife Federation, a group that later will promote preservation of rainforests as part of its many conservation efforts.

1944   Representatives of 44 nations attending a global conference establish the World Bank, which proceeds over the years to invest millions of dollars annually in development projects in nonindustrialized nations. In later years, environmentalists and

1944
(*cont.*) some political leaders will criticize the World Bank for funding projects that have adverse social and environmental impacts, although the bank will rescind some loans for projects proven damaging to the environment.

Because of dwindling supplies of timber on private lands and increasing demands for wood products nationwide, the U.S. Congress passes the Sustained Yield Management Act, allowing the Forest Service to sell timber from areas of national forests, providing the area can be renewed—reforested—in a specified period of time.

1946 The U.S. Bureau of Land Management is created to oversee public lands and maintain records of these lands.

1948 Leaders of national governments and nongovernmental organizations (NGOs) meet in Fontainebleau, France, to form the International Union for the Protection of Nature, which later will be renamed the International Union for Conservation of Nature and Natural Resources (IUCN). The organization will become an alliance of government agencies and NGOs representing 120 nations who try to find solutions to environmental problems and conservation of natural resources.

1949 Aldo Leopold's *A Sand Country Almanac* is published posthumously. The book presents Leopold's views on conservation in an essay titled "The Land Ethic," which declares that conservation is "a state of harmony between men and land." Leopold maintains as well that people should respect, admire, and love the order and integrity of natural communities—now called ecosystems. His book will be read by rainforest preservationists and other conservationists worldwide.

1950 At this time, 15 percent of the earth's land surface is covered with rainforest, but the rate of forest destruction is accelerating and within the century no more than 7 percent of the land surface is covered with forests.

1960 In response to increasing criticism over U.S. Forest Service management practices, the U.S. Congress passes and President Dwight Eisenhower signs the Multiple Use–Sustained Yield Act. The federal law requires the U.S. Forest Service to manage the renewable resources of the national forests so that "they are utilized in the combination that will best meet the needs of the American people" and so that there will be "high-level annual or regular periodic output of the various renewable resources

1960 of the national forests without impairment of the productivity
(cont.) of the land."

1961 Stewart Udall, U.S. Secretary of the Interior, articulates his views on the need for stewardship of the earth's resources. Udall serves until 1969, advocating an expanded national park system and writing *Quiet Crisis,* a book about the struggle to save the nation's natural resources.

1963 The U.S. Forest Service acquires Grey Towers, the former home of Gifford Pinchot, first chief of the USFS. The French chateauesque mansion in Milford, Pennsylvania, is surrounded by 102 acres of forest, formal gardens, and meadows. Grey Towers will become a National Historic Landmark where people are able to meet and discuss critical conservation issues, particularly those focusing on forest conservation.

1964 After 66 revisions, the U.S. Congress passes and President Lyndon Johnson signs the Wilderness Act, which creates a system to preserve wilderness areas, including temperate and tropical rainforests under U.S. jurisdiction.

1965 This year marks the first time that people from industrialized nations make contact with the Kayapo, an indigenous group in the Amazon who call themselves "people of the water's source." From this point on their lives and culture are increasingly under attack by invasions of loggers and others wanting to exploit the rainforest.

The United Nations sponsors a World Population Conference, calling attention to the link between rapid population growth in world population and environmental problems such as pollution and rainforest destruction.

1968 *The Population Bomb* by noted Stanford University Professor Paul Ehrlich is published, warning of global ecosystem deterioration, famine, and other major problems due to exploding populations. Ehrlich's book is one of a number of publications focusing on overpopulation issues over the next decade, bringing together groups calling for population control and environmental organizations. This and later books and articles by Ehrlich will help many in the environmental movement understand the connection between population growth and destruction of such resources as rainforests.

In Paris, France, representatives of international organizations meet in a conference called the Intergovernmental

1968
(*cont.*) Conference of Experts on the Scientific Basis for Rational Use and Conservation of the Resources of the Biosphere. The conference marks the beginning of a major international focus on environmental concerns.

A detailed plan to set up a protected park for one group of indigenous people, the Yanomami in Brazil, is presented to the nation's government, but the plan is ignored.

1969 The U.S. Congress passes the National Environmental Policy Act, incorporating some of the ethical standards in relation to the environment set by Aldo Leopold years earlier.

Because of concerns about the destruction of South America's forest Indians and their land, Survival International organizes in Great Britain to campaign for the rights of indigenous people.

1970 In Brazil, the government begins a colonization program known as Polonoroeste in largely inaccessible Rondônia and Matto Grosso in southwestern Brazil. Since the program is designed to resettle people from overpopulated areas of Brazil, a major highway, BR-364, is cut into the rainforest. Within ten years Rondônia's population will double and 20 percent of its rainforest will be gone.

The U.S. Congress creates the Environmental Protection Agency (EPA), which is charged with enforcing national standards for environmental quality.

U.S. political activists concerned about environmental issues establish the League of Conservation Voters. Their purpose is to campaign for, elect, and support, at all levels of government, public officials who sponsor or vote for legislation that protects the environment.

The first celebration of Earth Day, an idea developed by then U.S. Senator Gaylord Nelson, takes place on April 21 in cities, towns, and villages across the United States. At least 20 million people participate, setting the stage for the modern environmental movement in the United States, which incorporates rainforest preservation as a major part of its agenda.

1972 Representatives from 113 nations meet in Stockholm, Sweden, for a United Nations Conference on the Human Environment to discuss global environmental concerns. One result of the conference is the establishment of the United Nations Environment Programme, which will encourage and coordinate environmental programs within a variety of U.N. agencies.

1973   Congress passes the Endangered Species Act designed to prevent the loss of species. One provision of the law requires that the secretary of the Department of Interior in cooperation with the secretary of the Department of Commerce periodically prepare lists of endangered and threatened species for publication in the *Federal Register*. The law bans the export and killing of any animal or plant listed as endangered. No endangered species can be bought, sold, or taken from the wild without a special permit. Other provisions require that the habitats of endangered and threatened species, many of which are in rainforest areas, be protected.

1974   The U.S. Congress passes and President Gerald Ford signs the Sikes Act, which, among other provisions, requires federal agencies, such as the Forest Service and the Bureau of Land Management, to protect the habitats of endangered species and to initiate efforts to conserve fish and wildlife.

The Convention on International Trade in Endangered Species of Wild Fauna and Flora becomes an international law. The agreement is an attempt to control the illegal international trade in products of endangered species (such as ivory and fur).

1976   The U.S. Congress passes the National Forest Management Act, which requires each National Forest to prepare a management plan and sets limits on the size of an area that can be harvested by clear-cutting methods (cutting down all trees in large swaths of forest). The law allows timber cutting only on "such lands that can be restocked within five years after harvest." But the law does not state that reforestation is mandatory. Later, the law will be interpreted by the administration of President Ronald Reagan to mean that replanting does not have to take place in five years but must be possible within that time frame.

1979   James E. Lovelock's *Gaia: A New Look at Life on Earth* is published. Making use of the name that ancient Greeks gave to the Goddess of the Earth, *Gaia* is a theory about how the earth works—as a single living system with humankind only one small part of that system. Lovelock argues as well that life forms adapt to the earth and at the same time the planet adapts to life forms, regulating earth's climate and maintaining stability in spite of environmental changes. According to the Gaia theory, human activities, such as massive deforestation and fossil fuel use, are creating imbalances that could destroy life as we know it but would allow another form of life to emerge.

1980   A report on global environmental trends, commissioned by President Jimmy Carter in 1977, is published under the title "The Global 2000 Report to the President." The report describes serious environmental stresses worldwide and predicts that "life for most people on Earth will be more precarious in 2000 than it is now—unless the nations of the world act decisively to alter current trends." But Carter loses the 1980 election and the next administration of President Ronald Reagan ignores the report and begins to dismantle environmental regulations in the United States.

Thomas E. Lovejoy, a nationally known biologist, suggests an idea for rainforest preservation in nonindustrialized countries: swapping debt-for-nature conservation.

The U.S. Congress passes the Alaska National Interest Lands Conservation Act, which sets aside huge areas of federal land, including rainforest areas in Alaska, for preservation, recreation, and resource management.

1982   The United Nations Food and Agriculture Organization (FAO) publishes a study of worldwide deforestation, reporting tropical rainforest loss of 27 million acres (11 million hectares) annually. The report reveals that some reforestation is taking place, but there is still a net loss of 25 million acres (10 million hectares) each year. Yet these conjectures are based on outdated data, and satellite photographs and other studies will show later that the rate of forest loss is more than twice the FAO estimates.

1983   James Watt, appointed by President Ronald Reagan, resigns as U.S. Secretary of the Interior, primarily because of negative public reaction to his decisions to allow drilling for oil and gas and mining on public lands, including protected forestlands.

The United Nations establishes the World Commission on Environment and Development headed by Gro Harlem Brundtland, the Norwegian Labour Party leader who will later become Norway's Prime Minister. Made up of 22 policymakers and scientists from 22 countries, the commission sets out to develop proposals for action to deal with crucial problems in environment and development and to find ways to foster international cooperation.

1985   Rubber tappers from the Brazilian states of Rondônia, Acre, Amazonas, and Pará meet for the first time in Brasilia to discuss ways to preserve their forest homes and livelihood; they issue a manifesto on Amazonian development.

1987   After three years researching global development issues, the World Commission on Environment and Development publishes "The Brundtland Report" outlining strategy for global sustainable development—that is, development that does not destroy ecological systems. Among its seven goals is "merging environment and economics in decision-making" and assessing "potential impacts of new technologies before they are widely used, in order that their production, use and disposal do not overstress environmental resources."

Members of rainforest activist groups from around the world meet in Malaysia to organize a global movement to help save the world's rainforests. They also take part in the first annual Rainforest Week designed to focus attention on the need to stop the destruction of tropical rainforests.

1988   James Hansen, a scientist with the National Aeronautic and Space Administration Goddard Institute of Space Studies, tells a U.S. Senate committee that he is 99 percent certain that global warming is under way due to the buildup of greenhouse gases, which is partly caused by deforestation and the release of carbon dioxide from fallen trees, but is primarily the result of the massive fossil fuel use.

In December, Chico Mendes of Amazonia, who has organized rubber tappers working to preserve their way of life, is murdered in the state of Acre, Brazil.

1989   Jeff DeBonis, an American Forest Service employee (who resigns the following year), founds the Association of Forest Service Employees for Environmental Ethics (AFSEEE), an organization that provides a voice for workers who disagree with Forest Service practices that may have adverse effects on forest environments. The group, which totals over 2,000, believes the Forest Service has catered to business interests— selling timber—rather than managing forests, particularly ancient forests that are part of the coastal temperate rainforest, in an environmentally sensitive way.

Supported by major U.S. corporations, representatives of 250 U.S. groups, such as the American Mining Congress, the National Cattlemen's Association, and the western Environmental Trade Association, meet in Bellevue, Washington, to form a "Wise Use Movement," deliberately adopting the "wise use" term that Gifford Pinchot applied to argue against exploitation

1989
(cont.) of forests and other natural resources. But in contrast to conservation efforts, the coalition develops strategies to fight environmentalism with the same tactics used by grassroots organizations, even naming their individual groups with "environment friendly" terms such as Citizens for the Environment, Oregon Lands Coalition, People for the West!, and Information Council on the Environment. Among the coalition's agenda are antienvironmental campaigns demanding clear-cutting of ancient forests, the opening up of national parks and wilderness areas to mining and energy production, and the dismantling of the Endangered Species Act.

In April, a delegation that includes three Amazon tribal leaders travels from Brazil to London, England, to talk to the Prince of Wales and other influential people about the delegation's plan to extend and protect a Brazilian national park and Indian reserve in the Amazon.

In November, then Senator Albert Gore and 14 of his colleagues in the U.S. Congress meet with members of the European Parliament to form Global Legislators Organized for a Balanced Environment (GLOBE). The legislators plan to meet twice annually to exchange information on global environmental problems (including deforestation) and possible solutions.

1990 The government of Belize and Lighthawk (a U.S. environmental group known as "the air force of the environmental movement") announce the creation of the Bladen National Park, the first national park established in this small Central American nation. The new park will protect more than 80,000 acres of virgin tropical forest and will attract tourists and thus provide more jobs and income for the local people than would be possible through extensive logging operations. Prior to setting aside this protected reserve, the rainforest was threatened by proposals for logging concessions. Lighthawk helped promote the financial advantage of an intact tropical forest with its immense diversity of plants and wildlife.

Conservation International launches its Rain Forest Imperative, a ten-year strategy to sustain the richest and most critically threatened forests on earth. The organization's plan calls attention to 15 tropical forest "hotspots," three major tropical wilderness areas, and the two largest regions of temperate rainforest needing immediate protection.

1990   The U.S. Congress passes the Tongass Timber Reform Act
(cont.)   designed to eliminate mandatory cutting of timber to meet
targets set by the U.S. Forest Service and instead to cut only
enough timber in this temperate rainforest to meet realistic
market demands.

In June, a 36-year-old Englishman, Paul Coleman, a self-
described vagabond who has traveled extensively worldwide,
begins a 10,000 mile journey from Ontario, Canada, to the
Amazon rainforest, talking to hundreds of groups and news
reporters in his individual effort to call attention to the need to
preserve rainforests and to understand the people who live in
them. His journey ends two years later at Rio de Janeiro, Brazil,
in time for the United Nations' international conference on the
environment.

In July, more than a million Ecuadoran Indians stage a week-
long nonviolent protest in what would become known as the
Great Uprising, temporarily stopping expanded oil drilling on
a rainforest reserve. But the government retaliates by arrest-
ing and jailing many indigenous people and oil exploration
continues.

The northern spotted owl, whose habitat is ancient forests of
the Pacific Northwest in the United States, is listed as an endangered species.

1991   In April, a group of American activists begins a two-month
educational and cultural tour called a *chautaqua* to raise awareness and educate the public on the ecology of temperate rainforests. The chautaqua starts in British Columbia, Canada,
travels along the Pacific Northwest coast, and ends in San Diego,
California.

During the United Nations Environment Day in June, Claes
Nobel, great-grandnephew of Alfred Nobel who initiated the
prestigious Nobel Prizes, awards the first Nobel Prizes for environmental achievement in seven categories, including preservation of nature and natural resources.

In September, the tenth World Forestry Congress convenes in
Paris, a collaborative effort of France and the United Nations
Food and Agriculture Organization. The purpose of the con-
gress is to raise global public awareness of forest issues.

1991 On December 6, the government of Belize signs a statute creat-
(*cont.*) ing the Chiquibul National Park to protect lowland tropical
forest areas along the border with Guatemala.

1992 To protect sources of an anticancer chemical found in the bark
of yew trees, the U.S. Congress passes the Pacific Yew Act
designed to conserve yews in ancient forest stands, some of
which are part of coastal rainforests.

In January, a new coalition of 62 NGOs in the Amazon forms to
work toward preservation of Brazil's tropical forests. They
announce they will hold a series of regional meetings in the first
months of 1992 to consolidate their position and attract more
member groups.

British Columbia's New Democratic Party government an-
nounces in late January that it will ban logging for 18 months in
the temperate forests of Tsitika, Tashish, and Walbran Valleys.
Logging companies with cutting permits in these valleys will be
given timber to cut in nearby areas until a land use study is
completed. The government also says it will freeze logging
operations in three more areas in the province's interior.

In the United States, the Natural Resources Defense Council, a
nonprofit organization dedicated to protecting America's natural resources, cosponsors with the Yale Law School the first
major international conference on the relationship between
human rights and environmental protection.

In June, the Earth Summit convenes in Rio de Janeiro, Brazil.

The Brazilian government sets aside rainforest land for indigenous peoples, the Yanomami and the Kayapo.

1993 The United Nations declares this year as the International Year
of the World's Indigenous Peoples, in which policymakers and
an informed public worldwide attempt to set up guidelines to
establish land rights for indigenous people and resolve conflicts
over natural resources.

3

Biographical Sketches

PEOPLE WHO WORK TO PRESERVE tropical and temperate rainforests are from all walks of life and diverse areas of the world. Some are leaders of national environmental organizations and rainforest action groups, public officials, scientists, filmmakers, writers, and celebrities who have been recognized for their efforts to protect rainforests worldwide. Many people help preserve rainforests with their efforts to protect biodiversity, prevent species extinction, reduce population growth, and foster economic development that encourages sustainable use of natural resources. Thousands of activists who participate in organizations or make individual efforts will never be known except in their own groups or communities. Nevertheless, the rainforest movement and conservation of rainforest ecosystems could not survive without these unheralded volunteers.

Because of the vast number of activists and leaders worldwide whose activities directly or indirectly contribute to rainforest preservation, the profiles in this chapter are not meant to be comprehensive but rather representative. They include descriptions of grassroots activists as well as leaders of environmental groups, government officials, and business executives. A few of the sketches describe leaders from the past who helped pave the way for forest conservation and general environmental protection. Certainly some important contributors will be missing. But these profiles illustrate the diversity of involvement and the grassroots as well as global aspects of efforts to protect rainforests and their resources.

Edward Abbey

Called a radical environmentalist, Edward Abbey was a writer who today is best known for his 1974 novel *The Monkey-Wrench Gang,* which features characters who use wrenches and other tools to damage construction equipment and other machinery that destroy wilderness areas. Abbey's novel was a motivating factor in the 1980s Earth First! movement and its monkey-wrenching activities to protect ancient forests that make up much of the coastal temperate rainforests in the Pacific Northwest.

Abbey was born in 1927 in Home, Pennsylvania, and during the late 1940s and early 1950s was educated at the University of New Mexico and at the University of Edinburgh in Scotland. His 15-year tenure as a park ranger in the American Southwest inspired some of his writings. Besides his 1974 novel, Abbey wrote other works on the environment, including *Fire on the Mountain* (1962), *Desert Solitaire* (1968), and *Beyond the Wall* (1984). Abbey died in 1989.

James Neil Barnes

James Barnes is an expert on international environmental law and policy and is senior attorney and head of what was formerly the Environmental Policy Institute and is now the International Department at Friends of the Earth (FoE). His numerous responsibilities include overseeing FoE staff working on World Bank reforms and governmental policy changes to reduce tropical deforestation, climate change, and other global environmental problems.

Born in 1944, Barnes graduated in 1966 from Northwestern University in Illinois and received his law degree from the University of Michigan in 1970. For two decades he has specialized in international agreements on environmental protection, consumer rights, and human rights and has been a member of U.S. State Department negotiation teams developing international agreements on protection of the oceans and Antarctica. He served for six years as East Coast Director of Threshold International Center for Environmental Renewal and among other responsibilities served as advisor and policy analyst on tropical forest protection. In 1991, he received the International Environmentalist of the Year award from the National Wildlife Federation.

Brent Blackwelder

Brent Blackwelder is Vice President for Policy at Friends of the Earth United States and Treasurer of Friends of the Earth International. Blackwelder was educated at Duke and Yale Universities in the 1960s and received a Ph.D. in philosophy from the University of Maryland in 1975. For more than 20 years, he has worked to provide information for hundreds of independent citizen organizations nationwide that have been concerned about critical environmental and energy legislation. To carry out that effort, he established the Environmental Policy Center in 1972 and two years later set up the Environmental Policy Institute, which merged with Friends of the Earth and the Oceanic Society in 1988.

Among his many efforts to protect the environment, Blackwelder helped organize the global campaign to reform the lending practices of the World Bank and other development banks that were funding environmentally destructive projects. He helped to enact a number of laws requiring that development banks fund projects designed to protect tropical forests, promote energy conservation and renewable energy, and conserve biological diversity. At FoE he also set up staff to monitor federal appropriations in order to curtail state projects that destroy valuable natural resources, and to deal with environmental changes in the U.S. Tax Code.

In addition to his FoE duties, Blackwelder is a member of the Board of Directors of the League of Conservation Voters, the nation's largest environmental political action committee. He has appeared frequently before the U.S. Congress to testify on environmental policy and protective legislation and has discussed a range of environmental and energy issues on radio and television shows and in a variety of publications. With James Barnes and Walter Reid, he wrote *Bankrolling Successes: A Portfolio of Environmentally Sustainable Development Projects,* which was published in 1988.

David R. Brower

David Brower might be called the quintessential environmentalist. He is the founder of three major environmental organizations: Friends of the Earth, the League of Conservation Voters, and Earth Island Institute. He was the first paid executive director of the Sierra Club and has served on boards of numerous environmental organizations.

Born in 1912 in Berkeley, California, Brower has received honorary degrees from nine colleges and universities. In his younger years, he was an avid mountain climber and made dozens of first ascents. Known as an aggressive campaigner, he has been an environmental activist most of his life, organizing and leading conservation groups that protect rainforests worldwide. He also has arranged international conferences, bringing together various environmental, human rights, and peace groups to work on efforts to protect the Earth. He has edited and co-published numerous books for both the Sierra Club and Friends of the Earth.

Lester R. Brown

Lester Brown founded and is president of the internationally acclaimed Worldwatch Institute (WI), which tracks environmental trends. Brown, who was born in 1934, earned graduate degrees in agricultural economics at the University of Maryland and Harvard, and from 1959 to 1969, he served as administrator of the U.S. Department of Agriculture's International Agricultural Development Service, traveling extensively worldwide. He spent most of his travel time in rural areas where he observed the effects of environmental abuse—destruction of rainforests, overgrazing, soil erosion, desertification, and general deterioration of the agricultural base of many countries, a situation that he recognized would undermine these countries' economies. As the result of such observations, Brown began to monitor trends in environmental damage and founded WI in 1974 with a grant from the Rockefeller Brothers Fund.

Brown and the Worldwatch Institute have received numerous international environmental awards and prizes, and Brown speaks to groups around the world and counsels government leaders on policy changes that will help protect the planet. He calls for more fuel-efficient forms of transportation, much more recycling of goods made from nonrenewable resources, and most importantly a restructuring of the global economy so that the environment is sustained. He believes that environmental deterioration and social disintegration are linked and that national security demands conservation and wise management of the nation's and the world's natural resource base—of which the world's rainforests are an important part.

Jason W. Clay

Anthropologist Jason Clay is director of research at Cultural Survival, a nonprofit organization that advocates for the rights and welfare of indigenous peoples and tribal societies around the world. Clay also is founder and editor of the organization's magazine *Cultural Survival Quarterly*.

Born in 1951, Clay was educated at Harvard University, the London School of Economics, and Cornell University and holds a Ph.D. in anthropology. He has conducted research in Latin America and Africa on possible projects that could generate income for indigenous people while protecting biodiversity and has lectured extensively. He has written numerous materials—articles, papers, reports, and books—that focus on such topics as human rights and cultural survival of indigenous people.

In 1989, Clay established (and currently directs) Cultural Survival's marketing program, which generates income for forest residents who gather nontimber products, such as fruits, nuts, and oils, for sale to markets in North America and Europe. Funds from a rainforest benefit concert by the Grateful Dead held in New York in 1988 helped launch the program. Clay used the money to import more than 300 samples of rainforest products, eventually finding commercial markets for about 40 of them. In addition, Cultural Survival provided funds for a small nut oil processing plant in the Amazon, which is operated by the forest gatherers. Since 1989, Cultural Survival's marketing program has steadily grown, generating a total of $3 million in trade with 33 companies manufacturing 66 products by the end of 1991. Cultural Survival takes an agent's fee for the sale of the products, using those funds to support various programs, but net profits from product sales are returned to the countries of origin.

Barry Commoner

Ecologist and environmental activist Barry Commoner was called the "Paul Revere of the emerging science of survival" in 1970. Commoner, who was born in 1917, has long been active in conservation of natural resources such as forests. He is an elder of the U.S. environmental movement.

Educated at Columbia University and Harvard, where he earned a Ph.D. in 1941, Commoner taught for a short time at Queens College in New York, then served in the U.S. Navy during World War II. He returned to teaching after the war and during the 1980 election campaign was a Citizens Party candidate for President of the United States. He now directs the Center for the Biology of Natural Systems (CBNS), founded in 1966 at Queens College in New York. The center conducts interdisciplinary research on problems generated by modern technology and attempts to find solutions to environmental deterioration caused by various production and manufacturing methods. Commoner has written several books on the need for an integrated approach to life, emphasizing production methods that are in harmony with the environment and do not require destruction of rainforests and other ecosystems.

Jeff DeBonis

Jeff DeBonis is executive director of the Association of Forest Service Employees for Environmental Ethics (AFSEEE), which he founded in 1989 because of his concerns about the way U.S. forests were managed. A former USFS official who planned timber sales for the agency, DeBonis believes the Forest Service places the interests of the timber industry ahead of environmental values.

DeBonis was born in 1951 and grew up in the suburbs of Boston. He attended Colorado State University in Fort Collins, graduating from its forestry program in 1974. He then spent two years in El Salvador with the Peace Corps teaching soil conservation to farmers. In El Salvador, massive erosion of topsoil was the result of rainforest destruction due to slash-and-burn agriculture. Debonis later worked a short time for the Agency for International Development in Ecuador, then returned to the United States and took a job planning timber sales in the Kootenai National Forest in Montana where he first saw the effects of clear-cut logging—exactly the same situation he had seen in El Salvador. Rains falling on snow in the clear-cut areas caused sudden floods, sending soil from the bare hillsides sliding into creeks and destroying spawning sites for fish.

Although DeBonis was aware that wildlife biologists were concerned about ecological destruction in national forests, he at first accepted the Forest Service's aggressive timber-selling approach. Later he renounced their methods, especially after seeing devastation and soil erosion in the Willamette National Forest in Oregon. In 1991, he announced creation of AFSEEE and a quarterly publication, *Inner Voice,* that called for change within the Forest Service. DeBonis was invited to appear on television talk shows and before congressional hearings and groups of Forest Service employees across the United States. He quit the Forest Service a year after forming AFSEEE and became the organization's director.

Anne H. Ehrlich

Anne Ehrlich is a senior research associate in biology and policy coordinator of the Center for Conservation Biology at Stanford University. Since 1981, she has taught a course in environmental policy at Stanford. She has written or coauthored with her husband Paul Ehrlich numerous articles and books on population biology and has written extensively on such public issues as population control and how population growth affects the environment, particularly ecosystems such as rainforests. She is a consultant for many rainforest action groups.

Born in 1933 in Iowa, Ehrlich graduated from the University of Kansas in 1955. She began her career in the Department of Entomology at the University of Kansas, and in 1959 joined the research staff of Biological Sciences at Stanford, becoming associate director for the Center for Conservation Biology in 1987 and policy coordinator in 1992.

Anne and Paul Ehrlich frequently have pointed out in their writings, particularly in *The Population Explosion* and *Healing the Planet,* that the rapid increase in human population will have and is having a deleterious effect on the environment and the quality of human life. In a 1991 essay and lecture, the Ehrlichs warned that no nation can be secure if overpopulation leads to degradation and destruction of ecosystems. Soil fertility, plant pollination, nutrient recycling, a balance of gases in the atmosphere, biodiversity, and the hydrologic cycle are all vital to life on earth. According to the Ehrlichs, "the United States can be considered (in terms of its impact) the world's most overpopulated nation," because "the average American is a superconsumer, and the nation generally uses inefficient, environmentally damaging technologies," posing, on a per capita basis, a much greater threat to the planet than a poor nation exploiting its resources. (Source: Paul R. Ehrlich and Anne H. Ehrlich, "Population Growth and Environmental Security," *Georgia Review,* Summer 1991, 223–232.)

Paul R. Ehrlich

Paul Ehrlich is Bing Professor of Population Studies, Department of Biological Sciences, Stanford University, and has been a member of the faculty since 1959. The author of more than 500 scientific papers and magazine articles and 30 books mainly on ecology and evolution and the value of biodiversity, Ehrlich is widely known for his book *The Population Bomb* published in 1968. He is honorary president of Zero Population Growth, president of the American Institute of Biological Sciences, Fellow of the American Association for the Advancement of Science and the American Academy of Arts and Sciences, and a member of the U.S. National Academy of Sciences and the American Philosophical Society.

Born in 1932, Paul Ehrlich was educated in the public schools of Pennsylvania and New Jersey and received a graduate degree from the University of Pennsylvania in 1955. He earned a Ph.D. in biology at the University of Kansas in 1957, and has done extensive field, laboratory, and theoretical research in the field of population biology, which includes ecology, evolutionary biology, and behavior, collaborating with his wife Anne Ehrlich in policy research on human ecology.

Ehrlich's research has taken him to all continents, and his work is known worldwide, often serving as a resource for people working on rainforest preservation projects. He has received numerous awards, been a frequent guest on television and radio programs, and has given hundreds of public lectures on ecology, human ecology, and evolution. Through these efforts he has consistently promoted ways governments and individuals can make changes that will help sustain life on the planet.

Jerry Franklin

Jerry Forest Franklin is chief plant ecologist at the U.S. Forest Service Pacific Northwest Research Station and Bloedel Professor of Ecosystem Analysis at the University of Washington. Franklin is known for his "New Forestry" method of resource management, which he calls "a responsible and holistic view of the forest that sacrifices no one resource to another."

Franklin grew up in Camas, Washington, a pulp mill town, and received his degree in forestry at Oregon State. In 1969, Franklin participated in the International Biological Project, a United Nations–sponsored study of the major ecosystems of the earth. He led a group of scientists studying the old-growth forests in the Pacific Northwest, conducting research on the ecology. At that time, the conventional view in forestry was that trees were commodities similar to staples like corn and wheat crops, and since old-growth trees had "matured" and would not increase in size, they should be harvested and replaced with fast-growing seedlings. But the study of Pacific Northwest forests showed that ancient forests were part of a complex rainforest ecosystem, with many hundreds of species of plants and animals linked and dependent on this particular habitat. The findings led Franklin to further research and eventually to coauthoring many technical papers on ecology and a textbook on the biology of forests in the Pacific Northwest.

Franklin's new forestry approach to timber management is to log in a deliberately messy fashion, leaving snags, fallen logs, and other woody debris to provide homes for wildlife. Although there are many critics of new forestry techniques, Franklin continues to speak out, write, and press for a forest management strategy that takes into account the entire ecosystem, maintains diversity, and allows logged areas like those in the coastal rainforests to recover. At present the U.S. Forest Service is experimenting with Franklin's concept, but only on a limited basis.

Lou Gold

"As an Oregonian I have taken my stand on Bald Mountain, but wherever you live, these lands belong to you as a citizen and a taxpayer. As a part owner, you have a right and perhaps an obligation to make your voice heard in the debate over how our national lands should be managed." These are the words of Lou Gold in an article for the Winter 1990 edition of *Orion* magazine. Lou Gold believes strongly in the idea that one person can make a difference.

In middle age, Gold gave up a successful university teaching career in Illinois after participating in an Earth First! protest against construction of a logging road in a wilderness area of the Siskiyou mountains in southern Oregon. Gold moved to Oregon and since 1983 has spent his summers as a self-appointed guardian of Bald Mountain in the Siskiyou range, the most rugged in the Pacific Northwest. Gold maintains a solitary camp and keeps a vigil atop the mountain, watching for signs that the U.S. Forest Service has allowed logging or further road construction into the wilderness. He conducts what he describes as a "medicine-wheel" ceremony every evening, a partly fanciful, partly serious ritual and prayer service to save a forest from USFS timber sales.

In the winter months, Gold travels across the United States presenting a slide show and explaining the need to protect the biologically rich ancient forests that are part of the temperate rainforests in the Pacific Northwest. He emphasizes that people in the United States have no right to criticize deforestation in distant places like the Amazon when they continue to condone logging and road building in the last remaining pristine forests at home.

Randall Hayes

Called an "action-oriented conservationist," Randall Hayes is founder and director of the Rainforest Action Network. He was born in 1950, and his training ground as an activist was documentary filmmaking, producing the award-winning film *The Four Corners, A National Sacrifice Area?* This film documents the tragic effects of uranium and coal mining on Hopi and Navajo lands in the American Southwest. While working with Indian groups on the film, Hayes learned about the plight of native people in tropical rainforests.

As director of the Rainforest Action Network, Hayes is a leader in the effort to halt destruction of tropical rainforests and the fight to protect the rights of indigenous people. He has written and lectured extensively on tropical rainforests and activism. The lecture and slide program that Hayes presents emphasizes that rainforest devastation is one of the major threats to human life. Slides show the importance of rainforest products to everyday life, and some of the causes and effects of destruction, particularly overconsumption and exploitation of rainforest lands.

However, Hayes also points out ways that activists can make a difference. He works with organizers and regional networks throughout North America, Latin America, Africa, Asia, and Europe. His efforts have helped to demonstrate how a network of activists can help protect rainforests worldwide.

Andy Kerr

Born in 1955 and a fifth-generation Oregonian, Andy Kerr is Conservation Director for the Oregon Natural Resources Council (ONRC). He has worked with the organization since 1976, starting while in his mid-twenties as a grassroots field organizer. Kerr now is responsible for ONRC's conservation program and works extensively with both the Oregon legislature and the U.S. Congress, seeking better legislative protection for Oregon's natural resources.

Kerr has been characterized as the timber industry's "most hated man in Oregon" and a "white collar terrorist" because of his effective efforts to prevent damaging exploitation of Oregon's wilderness. He has helped to curtail temporarily logging operations in rainforest areas of the Pacific Northwest by filing legal appeals against the U.S. Forest Service's planned timber sales. Kerr has appeared numerous times on network television news and feature programs on forest and other natural resource issues. He also lectures frequently at universities and speaks to a variety of commercial, professional, and trade associations about the need for forest conservation, and is contributing editor of *Forest Watch* magazine.

Kerr's ability to call attention to ancient forests has resulted in increases in ONRC membership and funds and greater public awareness of the need for temperate rainforest protection. Currently, Kerr oversees ONRC campaigns that include efforts to protect the state's remaining ancient forest stands, establish new National Parks in Oregon, and ban offshore oil exploitation along the Oregon Coast.

Roxanne Kremer

Executive director and cofounder of the International Society for the Preservation of the Tropical Rainforest, Roxanne Kremer is a naturalist living in Southern California. Kremer grew up in rural Wisconsin, where she gained an understanding of the interrelatedness of all living things. She bases her conservation efforts on a premise that she often states: "We are all interconnected with one another as if one beating cell. Kill a part of that cell and the whole cell eventually dies."

Since the early 1980s, she has worked extensively on an internationally renowned project called Preservation of the Amazonian River Dolphin (PARD) to protect the "pink" dolphin, considered to be the most intelligent of all dolphin species. The project is patterned after a successful gorilla conservation program that Kremer established in Rwanda, Africa, before going to the Amazon. Called the Educational Conservation Tourism program, the Rwanda project helps conserve gorillas by demonstrating to the local people that the animals are better alive than dead. Previously village people had killed gorillas, cutting off their hands to sell as ashtrays. Villagers found the live gorillas an important source of revenue because tourists want to see the animals in their natural habitats. Kremer and PARD hope that the same kind of incentives will help save the Amazonian River dolphins.

Aldo Leopold

Born in 1887, Aldo Leopold was a pioneer environmentalist and nature writer known for *A Sand Country Almanac*.

Leopold earned a master's degree in forestry at Yale University, then in 1912 joined the U.S. Forest Service, supervising the Carson National Forest in New Mexico, where he became convinced that some of the area should be left undisturbed by roads and preserved in a wild state. After two years in an administrative position with the USFS's Forest Products Laboratory in Wisconsin, Leopold left the agency and became a forestry and wildlife consultant and eventually wrote and taught on the subject. In 1935, he helped organize The Wilderness Society, an organization that has conducted many programs to preserve rainforests. He died in 1948, and a year later his *Almanac* was published. The book called for a "land ethic" in which people viewed land as part of their community not as a commodity to own and exploit.

Norman Lippman

Filmmaker and plant scientist Norman Lippman is director of a nonprofit Documentary Project, a work-in-progress series of documentary films on managing tropical rainforest areas sustainably. The series titled *Talking with the God of Money* will show sustainable and profitable alternatives to rainforest destruction, how four communities of forest peoples in Latin America manage their forest resources, and how these communities may adapt to a market economy. The films will also help inform the public and policymakers on the value and potential of rainforests and their inhabitants and identify the causes and regional and global impact of rainforest destruction. A basic theme and message of the entire series is that indigenous people must benefit from the rainforest in order to ensure the protection and sustainable management of these vital resources.

Lippman, who was born in 1952 and grew up in St. Louis, studied film and communication arts at Columbia University in New York City and Video Field Production at the North American Television Institute in White Plains, New York. He also studied plant science at the Center for Tropical Agriculture at the University of Florida.

While studying plant science, Lippman worked for two years in the Guatemalan tropical rainforests and highlands directing two grassroots agricultural development programs. Later, he spent 17 months living with and studying the Lacandon Maya Indians in southern Mexico, sharing in their work and rituals and filming their resource management cycle for an award-winning film, *Keepers of the Forest*. The film was shown periodically in 1992 on the Discovery Channel and is available for purchase and rental.

The current documentary project began in 1986 and is being developed in affiliation with the Missouri Botanical Garden and the St. Louis Ambassadors Foundation. It is supported by donations and major funding from the John D. and Catherine T. MacArthur Foundation. Like *Keepers of the Forest*, the new series will include footage of the Lacandon Maya way of life. In-depth interviews with Indians, settlers, government officials, and scientists are also part of the series. Initial broadcasts of the documentaries will be in English, Spanish, and Portuguese, and plans are to distribute various versions of the series to secondary schools and universities along with appropriate educational materials.

Thomas Eugene Lovejoy

A biologist, conservationist, and leading Amazon researcher, Thomas Lovejoy is assistant secretary for external affairs of the Smithsonian Institution. He is considered one of the world's leading environmentalists and serves on a variety of boards of wildlife conservation organizations. He has been an administrator of numerous conservation and research institutions.

Lovejoy was born in 1941 and was educated at Yale University, receiving a Ph.D. in biology in 1971. During most of his career, he has called for action to protect the rapidly disappearing tropical rainforests and the species that are destroyed with them.

In the early 1980s, Lovejoy articulated an idea for rainforest preservation: debt-for-nature swaps. Following such a strategy, an environmental group raises funds for tropical nations to repay debts owed to international development banks. In exchange, the debtor nation pays back its debt to the environmental group, which uses the interest on the loan to buy tracts of rainforest land or to pay for local projects designed to preserve forests. In the United States, legislation was passed in 1987 allowing various environmental groups to act upon Lovejoy's idea.

José Lutzenberger

A renowned ecologist, José Lutzenberger is director of the Brazilian chapter of the Gaia Foundation in Porto Alegre. The foundation provides funds for research in sustainable agriculture and supports various projects by indigenous people and rubber tappers who are helping to preserve the Amazon rainforest by harvesting products rather than destroying trees.

Lutzenberger was the first minister of environment for Brazil, appointed in 1990, but was fired from the position in 1992 because, according to official accounts, he had become "too radical" and opposed any type of development in Amazonia. But Lutzenberger has long been opposed to massive destruction of rainforests in Brazil and other parts of the world and has been an outspoken critic of development practices that do not sustain natural resources. He has often called modern industrial society "a cancer" on the environment, and Brazilian industrialists, landowners, and military leaders see him as a formidable opponent.

Born in Brazil in 1927, Lutzenberger studied agronomy during the 1950s. For 15 years he worked in Germany for a chemical company, concentrating on fertilizers and pesticides. According to a profile by Andrew Revkin in *Audubon* magazine (November/December 1991), Luztenberger returned to Brazil in 1971 and "found that pesticides and development had poisoned and transformed the landscape. The wetlands where he had hiked as a student and land surveyor were sterile and silent. . . . " Because of the changes, he became an uncompromising environmentalist and activist.

George Perkins Marsh

Architect, linguist, politician, diplomat, geographer, scholar—these are just a few of the terms used to describe George Perkins Marsh, called the "original environmentalist" by some conservationists today. Born in 1801, Marsh became an expert in diverse fields, but of his many accomplishments, the one most recognized is his scholarly book *Man and Nature: Or, Physical Geography as Modified by Human Action,* published in 1864.

In *Man and Nature,* republished in 1965 by Harvard University Press, Marsh spelled out his theory that degradation of the natural environment led to the decline of earlier civilizations. Marsh had seen in his home state of Vermont and in countries around the world the effects of deforestation on soil, plant and animal life, and streams. He predicted that the earth would become unfit for human habitation if people continued to destroy natural resources, which in turn would lead to impoverishment, crime, barbarism, and "perhaps even extinction of the species."

Marsh died in 1882 but today is remembered as one of the first to write about the interrelatedness of all life, describing the concept of the web of life and what is now called the science of ecology. He emphasized that nature by itself could not necessarily restore or heal the destructive effects of human activities—that people needed to aid in the restoration process.

Chico Mendes

Francisco "Chico" Alves Mendes Filho, known as Chico Mendes, was born in 1944 in the partially deforested ranchland of Porto Rico, Brazil. His family worked and lived on a rubber plantation, growing their own food or harvesting it from the nearby forest. Chico, the oldest of six children (out of seventeen who survived), learned early how to tap rubber trees and collect the latex. He also learned in early childhood how rubber barons and other wealthy landowners and military leaders in Brazil had long suppressed and exploited peasants, indigenous Indian populations, and poor immigrants. By the time he was in his early twenties, he was trying to organize rubber tappers to demand schools, better working conditions, and fair prices for the goods they had to buy from company stores.

Mendes left the rubber plantation in 1971 and taught adults at a government school. About the same time, the Brazilian government was building roads through the rainforest to the state of Acre where the rich rainforests are located, offering incentives to attract industry and cattle ranchers. Mendes saw for himself hired workers burn and cut down forests, driving rubber tappers and poor farmers from the land.

Mendes, with the help of activist Catholic priests, joined several other grassroots leaders and began to teach rubber tappers their rights and to organize unions, a dangerous undertaking since many leaders and members were killed or beaten by ranchers or their hired thugs. By the early 1980s, Mendes was actively recruiting unionists and working with a national organization of rural workers who held a conference in Brasilia in 1984, bringing together more than 4,000 delegates. During the conference,

Mendes promoted a forest region plan that would eventually become the basis for extractive reserves—regions set aside to preserve the forest. Although his plan was not adopted that year, Mendes continued his fight—in spite of threats on his life and the murder of other activists, including rural workers, priests, and lawyers.

In 1987, Mendes received the United Nation's Global 500 award and the Better World Society medal for environmental conservation. His efforts also were featured in a major New York *Times* article and in other major newspapers and magazines in the United States. But Mendes was in constant danger from enemies in his own country who were enraged because of his success in preventing rainforest destruction, and in December 1988 he was murdered by cattle ranchers. His death brought international attention to the Amazon rainforest and the efforts of many brave people to save it. *The Burning Season* by Andrew Revkin and *The World Is Burning: Murder in the Rain Forest* by Alex Shoumatoff tell the story of Chico Mendes and the fight for the Amazon rainforest.

George Miller III

U.S. Representative George Miller III (D-CA) is chairman of the Interior and Insular Affairs Committee, taking over the seat long held by Morris Udall, who retired in 1991. Miller is known as a tough environmentalist in Congress and for more than a decade has crusaded for such issues as providing fair water allotments in California, reducing the use of toxic pesticides in agriculture, and protecting the coastal rainforests in the Pacific Northwest.

Born in 1945, Miller grew up in the San Francisco Bay area and learned about politics from his father, a state senator in California. After working on tugboats and in the merchant marine, he earned a law degree at the University of California, Davis. In 1974, he won election to the U.S. Congress and became a member of the Interior Committee. Now as chairman, he fights for protection of public lands, with special emphasis on coastal rainforests, from over-exploitation by private interests, such as mining and timber companies.

Norman Myers

A leading British ecologist and an international consultant in environment and development, Norman Myers focuses his research on resource relationships between the developed and developing countries. A winner of the World Wildlife Fund gold medal, Myers has been a consultant to various United Nations agencies, the World Bank, the U.S. National Academy of Sciences, the Smithsonian Institution, and many other organizations and has analyzed the problems of deforestation in 60 tropical countries.

Myers has written numerous magazine articles and books on rainforest issues. Among the books he has written or edited are: *A Wealth of Wild Species: Storehouse for Human Welfare, Conversion of Tropical Moist Forests, The Sinking Ark, The Primary Source: Tropical Forests and Our Future,* and *The Gaia Atlas of Planet Management.*

Gaylord Nelson

Frequently called the "Father of Earth Day," Gaylord Nelson for decades has been a staunch environmentalist and conservationist. He became Counselor of The Wilderness Society in 1981.

Born in 1916, Nelson was educated at San Jose State College in California and earned a law degree at the University of Wisconsin Law School. He served for 46 months in the U.S. Army during World War II, and after the war practiced law for 12 years, beginning a political career as a state senator during that time.

Nelson served a four-year term as governor of Wisconsin and was elected to the U.S. Senate in 1962, serving for 18 years. While in the Senate he sponsored numerous bills, including the 1964 Wilderness Act, for protection of the nation's forests and other natural resources. In 1970, he founded Earth Day as an annual observance focusing national (and later global) attention on such environmental problems as destruction of rainforests. He was honorary co-chairman of Earth Day 1990.

Evaristo Nugkuag

A member of the Aguaruna Indian tribe, Evaristo Nugkuag is a grassroots organizer in Peru. Born in 1950, he was educated at missionary schools and like Chico Mendes saw as he was growing up how outsiders were destroying his homeland. In 1981, he brought together 13 tribal groups representing more than half of the indigenous people of Peru's rainforest to find ways to protect basic human rights of indigenous people and the traditional Indian lands from the encroachment of timber and mining companies as well as from drug dealers.

In 1984, Evaristo Nugkuag negotiated with tribal leaders in other nations to develop the Coordinating Organization for

Indigenous Bodies in the Amazon Basin (COICA). COICA now represents more than half of the indigenous people in Amazonia, including tribal groups not only in Peru but also in Colombia, Brazil, Ecuador, and Bolivia. The organization works to defend Indian rights, land, and resources and to bring about political representation and education for indigenous people.

As a grassroots leader, Evaristo Nugkuag has received several international awards, one of which was the 1991 Goldman Prize, created the year before by the Goldman Environmental Foundation set up in the United States by a wealthy San Francisco couple who wanted to call attention to and annually award grassroots leaders on six continents for their "efforts to preserve and enhance the environment."

Randal O'Toole

Born in 1952, Randal O'Toole earned degrees in forest management and geology and completed three years of graduate work in economics. He founded Cascade Holistic Economic Consultants (CHEC) of Portland, Oregon, in 1975. O'Toole established the organization to provide educational and technical services to the environmental movement, particularly in relation to public forests, a great portion of which are rainforests. Today CHEC advocates free market solutions to many other environmental controversies.

Under O'Toole's direction, CHEC was one of the first conservation groups to use a sophisticated economic model to argue for preservation of wilderness areas, which include temperate rainforests. In 1982, he began studying below-cost timber sales from U.S. forests and analyzing dozens of U.S. Forest Service plans, which led the USFS to greatly reduce timber sales and protect millions of acres of wildlands. Based on his research, O'Toole wrote *Reforming the Forest Service,* which was published by Island Press in 1988.

Paulo Paiakan

Paulo Paiakan, who has no record of his birth date but probably was born about 1954 or 1955, is a chief of a Kayapo tribe in the Amazon rainforest and has brought together indigenous people and environmental groups to work cooperatively in efforts to save the forests and the people who live there. Descended from a long line of chiefs, Paiakan knew early in his life that his special destiny was to "go out into the world to learn what was coming" to his people, as he told a *Parade Magazine* journalist.

During his teenage years Paiakan saw the effects of road building—the destruction of forestland to construct the Trans-Amazonian Highway, and additional devastation brought by loggers, ranchers, and miners who followed. The chief tried to convince villagers to leave their homes and move farther into the forest where they would be able to maintain their way of life. Most villagers thought the forest could not be destroyed, and only about 150 people agreed to move away, finally settling in Aukre.

Although the Aukre villagers found a better life, many other Kayapo were adversely affected by pollution and disease that came with the invaders. Paiakan, who had attended a missionary school for his early education, decided to fight for his people and during the 1980s went to the state capital, Belem, where he learned to live like white people and to speak their language (Portuguese). His mission was to educate government leaders about threats to indigenous people and to convince Indians to protect their forest-lands and to ward off temptations to trade their timber for flashy manufactured goods.

Over the decade, Paiakan learned to operate a video camera so he could film the deforestation activities and show them to his people; he organized protests against a planned hydroelectric dam that would flood vast areas of rainforest; he traveled to the United States, Europe, and Japan to publicize the threats to Amazon Indians. In the early 1990s, he began efforts to convince Indian leaders that selling rainforest products was a way to preserve the forest and earn income. Paiakan also organized indigenous leaders, making arrangements for them to attend a conference of NGOs held at the same time as the Earth Summit in Rio de Janeiro in June 1992. His work to maintain the Kayapo culture and at the same time convince the outside world to work with nature rather than overcome it goes on. (For a more detailed account of Paulo Paiakan's efforts, see Hank Whittemore's article "'I Fight for Our Future,'" in *Parade Magazine*, April 12, 1992, pp. 4-7.)

Gifford Pinchot

Gifford Pinchot served as the first chief of the U.S. Forest Service. Born in Connecticut in 1865, Pinchot was a member of a wealthy family of merchants, politicians, and landowners and was educated at some of the best eastern schools. His father urged him to make forestry his profession, but no U.S. university offered such a degree at that time. After graduating from Yale, Pinchot studied forestry in France, returning to the United States to work as a forester and later in 1896 to become part of the National Forest Commission set up to identify forest areas for possible forest reserves (now known as national forests, some of which are rainforests).

In 1898, Pinchot became head of the Division of Forestry in the Department of Interior. When President Theodore Roosevelt took office, the management of forest reserves was transferred from the Department of the Interior to the Department of Agriculture, and Roosevelt named his friend Pinchot Chief Forester. With Roosevelt's support, Pinchot restructured and professionalized the management of the national forests. Until the early 1900s, there was no system in the country for managing private or government-owned timberland. As Pinchot observed, "to waste timber was a virtue not a crime." Lumbermen "regarded forest devastation as normal and second growth as a delusion of fools."

Generally considered the "father" of U.S. conservation because of his unrelenting concern for the protection of the U.S. forests, Pinchot emphasized long-term forest management and wise use of timber resources. Under Pinchot's guidance as forest chief from 1905 to 1910, the number of national forests more than doubled. Pinchot also founded the Society of American Foresters and the School of Forestry at Yale University and wrote numerous reports and books on conservation as well as an autobiography, *Breaking New Ground,* which has been reprinted several times and explains his views of the U.S. conservation movement. Pinchot died in 1946.

Ghillean T. Prance

Director of the Royal Botanical Gardens outside London in Great Britain, Ghillean Prance has long been an environmental activist and has a deep reverence for the earth, in particular tropical rainforests. He has worked extensively to demonstrate that rainforests should be saved because they provide economic benefits for a nation.

Prance grew up on the Isle of Skye off the coast of Scotland and was educated at Oxford, earning three degrees. For 25 years he was a staff member of the New York Botanical Gardens and frequently spent time in Brazil carrying out extensive botanical explorations. While in New York, he established the Economic Botany Institute to study the potential yield and value of products from tropical forests that are managed wisely. He is considered one of the foremost authorities on the Brazilian rainforest and is the author of eight books and nearly 200 articles on rainforest botany.

Barbara Y. E. Pyle

Barbara Pyle is Vice President of Environmental Policy for Turner Broadcasting System (TBS), Executive Producer of the International Documentary Unit, and CNN's Environmental Editor. She supervises the development of environmental programming and designs the environmental policy for TBS. She also oversees the entire production of issue-oriented documentaries and CNN's newsgathering efforts on environmental stories worldwide and advises the daily environmental story called "Earth Matters," which she originated.

Pyle, who was born in 1946 in Oklahoma, was educated at King's College, England, and Tulane University where she was named the Outstanding Student of Philosophy and a Woodrow Wilson Fellow. She earned a master's degree in philosophy and logic at New York University. She has lived and worked in numerous countries and has been a photojournalist and aerial photographer; many of her photographs have appeared in major consumer magazines.

In 1980, Pyle joined TBS to make critical global issues understandable and available to the widest possible audience. Her past productions have covered a variety of environmental concerns, with special emphasis on deforestation in the tropics and the threats to indigenous people. Her films have won more than 50 awards. Currently she is co-executive producer for *Captain Planet and the Planeteers,* an animated action-adventure series about a superhero and five children battling environmental problems.

At the 1992 Earth Summit, Pyle represented all global broadcast media and originated the Save the Earth campaign, a series of programs and other activities focusing on the Earth Summit. A popular speaker, Pyle has addressed a variety of environmental conferences and educational and social organizations and has been honored for her outstanding contributions to the protection and improvement of the environment.

Peter H. Raven

Director of the Missouri Botanical Garden in St. Louis, Peter Raven is also Engelmann Professor of Botany, Washington University, St. Louis, and Adjunct Professor of Biology at St. Louis University and the University of Missouri–St. Louis. He is a leading authority on tropical forests and their inhabitants, and his work is known worldwide. In his position at the Missouri Botanical Garden, he has employed over 30 botanists and other specialists who are assigned to work throughout the tropics, researching and cataloging new tropical plants. Specimens are collected and brought to the St. Louis herbarium for further study.

Born in China in 1936, Raven was educated at the University of California, Berkeley, and the University of California, Los Angeles. He holds honorary degrees from numerous universities and has received dozens of awards and honors from scientific, botanical, horticultural, and conservation organizations. He also has written hundreds of scientific papers on tropical plants and animals and has served on numerous editorial boards. Raven has coauthored or coedited at least 16 books, including *Biology of Plants, Understanding Biology, Modern Aspects of Species, Research Priorities in Tropical Biology,* and *Coevolution of Animals and Plants.*

Anita Roddick

"THE INDIANS ARE THE CUSTODIANS OF THE RAINFORESTS. THE RAINFORESTS ARE THE LUNGS OF THE WORLD. IF THEY DIE, WE ALL DIE." These words are typical of the kinds of slogans printed on semi trucks that transport products from The Body Shop International, a cosmetic manufacturer and retailer founded by Anita Roddick in Littlehampton, England. Roddick started the business in 1976 with several goals in mind: to produce and sell "natural" skin and hair care products, to foster education, and to promote the concept of service to others. Rather than using the usual marketing techniques, the company, which began in a storefront, has depended on word-of-mouth advertising, store window campaigns, community outreach programs, and mail order catalogs to reach its customers. Expanding rapidly, the company now includes more than 800 shops in 40 countries around the world, including Canada and the United States.

To Roddick, principles are as important as profits, so she and her staff spend a lot of time on quality-of-life issues, such as environmental concerns and human rights, actively supporting groups like Cultural Survival and various international environmental organizations. The Body Shop is as well known for its environmental activism as for its products. Roddick and her company also promote the concept of "trade not aid" as a positive approach to solving economic problems in the developing world. Rather than giving handouts, the company helps communities around the world obtain the resources they need to support themselves. Trade-not-aid projects have helped indigenous groups in the tropics develop and market sustainable forest products.

John Seed

Born in Budapest, Hungary, in 1945, John Seed today is known as the "town crier of the global village" and is a leading activist for rainforest protection. Seed makes his home in Australia but has traveled worldwide, lecturing and showing films to raise awareness and prompt action on behalf of the world's threatened rainforests and endangered forest people and species. He founded the Rainforest Information Centre in 1979 after learning about the threats to Australia's rainforests and taking part in nonviolent but direct confrontation tactics to save these resources. His activism has included everything from burying himself in front of bulldozers to door-to-door political campaigns that have helped prevent rainforest destruction.

Seed earned his college degree in psychology and philosophy from Sydney University, Australia. Although not a biologist, he has studied rainforest issues extensively and has become so well informed that many scientists depend on his information bulletins to keep them updated on rainforest issues. He writes and lectures frequently on deep ecology and conducts Council of All Being retreats/rituals in Australia and North America, raising consciousness about the environment and spreading the philosophy of deep ecology that teaches the interconnectedness of all life.

Michael M. Stewartt

Michael M. Stewartt is founder and president of an unusual environmental organization called Project Lighthawk, an "environmental air force" based in Santa Fe, New Mexico. A professional pilot and amateur biologist, Stewartt began Lighthawk because he believed that an aerial view of environmental degradation was far more convincing than pages of facts and figures attempting to prove the same thing.

Born in Tucson, Arizona, Stewartt spent his early years exploring the majestic southwestern desert of the United States. He worked as a wilderness guide/river runner in the American Southwest and also in Canada, spent two years as a bush pilot in Nome, Alaska, and flew as a volunteer for Wings of Hope, providing aid for people in Guatemala. He also has been a commuter airline pilot.

In 1974, Stewartt learned that an environmental group in Santa Fe, New Mexico, was fighting the construction of a coal-fired power plant in southern Utah because smoke and other air pollutants from the plant would hamper if not obscure views of the nearby Grand Canyon. Stewartt volunteered to fly TV camera crews and photojournalists over the area to see for themselves where the proposed plant would be. News media pictures of the site prompted public outrage and officials of the power company eventually dropped the construction project.

The see-for-yourself concept is the basis for Lighthawk's services, which are used by many environmental groups trying to protect temperate or tropical rainforests. Stewartt and his Lighthawk Project have been instrumental in helping to shape public opinion about the need for conservation of rainforests and other natural resources. In 1988, the National Wildlife Federation honored Lighthawk with the Conservation Organization of the Year award, and in 1990 Stewartt received Chevron's Conservation Award.

Frank Wadsworth

Called "Mr. Tropical Forestry," Frank Wadsworth has worked as a forester and researcher in tropical forests for 55 years, beginning long before such issues as population explosion in the tropics and destruction of rainforests were publicized in the United States. Although not all tropical forests are rainforests, much of Wadsworth's work has been in rainforest areas. He has served as advisor to tropical forestry projects in Mexico, in Central and South American countries, and in Kenya, Africa, and in Sarawak, Malaysia.

Born in 1915, Wadsworth was raised in Chicago, Illinois, but during his early years became interested in forestry because of his frequent bird-watching forays with friends through the forest preserves around the city and metropolitan area. He also gained appreciation for trees and forests while traveling with his father, who was in the paper business, to paper mills located amidst the forests of Wisconsin and Minnesota. He earned his Ph.D. in forestry at the University of Michigan and began his Forest Service career in Arizona, researching ponderosa pine.

In 1956, Wadsworth became director of the USFS Institute of Tropical Forestry in Puerto Rico, researching on and teaching others about sustainable forest management. He has written numerous articles and several books on forestry. At present the institute is expanding, and Wadsworth is now acting as an assistant director for international cooperation, a USFS effort to arrange cooperative agreements on forest management with forest service institutions in tropical countries.

Edward O. Wilson

A leading U.S. entomologist, Edward O. Wilson is Professor of Zoology and Curator in Entomology at the Museum of Comparative Zoology at Harvard University. He is known worldwide for his work in sociobiology—the biological basis of social behavior—and is acclaimed for his expertise on insect societies. His many books include *The Insect Societies* (1971), *Sociobiology: The New Synthesis* (1975), the Pulitzer Prize-winning book *On Human Nature* (1979), *Biophilia* (1984), and *The Diversity of Life* (1992). He also edited *Biodiversity* (1988), a compilation of papers defining many rainforest issues.

Born in Birmingham, Alabama, in 1926, Wilson spent much of his childhood in the rural areas of northern Florida and adjacent Alabama counties where he not only roamed the woods and enjoyed hunting and fishing, but also, in his words, "cherished natural history for its own sake and decided very early to become a biologist." Wilson was educated at the University of Alabama and Harvard University and was appointed to the Harvard faculty in 1956.

Wilson has conducted research in rainforests around the world, and has received numerous awards from scientific organizations in the United States, Sweden, France, and other countries. He is a member of the Club of the Earth made up of nine prominent U.S. scientists who believe that species extinction of unimaginable magnitude will occur if broad conservation efforts are not undertaken in the near future.

4

Facts and Statistics

IN GENERAL, THIS CHAPTER supplements and illustrates information provided in Chapter 1. It is divided into 5 sections: (1)
general misconceptions about rainforests; (2) data on temperate
rainforests, including a map showing the locations of coastal rainforests in North America and tables and graphs depicting logging
practices in these rainforests; (3) data on tropical rainforests
worldwide, including lists of products from the forests and graphs
showing deforestation in selected areas of the world; (4) graphs
showing carbon dioxide buildup in the atmosphere, due in part to
deforestation; and (5) documents on rainforests prepared by
national and international organizations.

Misconceptions about Rainforests

Misconceptions about rainforests abound. Examples of some of
the most common myths are shown in Table 1 (see next page).

Temperate Rainforests

Many North Americans are just becoming aware of and beginning
to learn about rainforests on their own continent. This section
illustrates where coastal rainforests are located, the amount of

TABLE 1   Rainforest Myths and Facts

Myth: Even though areas of rainforest are cut down or burned, the forest will grow back in time or trees can be replanted, creating another rainforest.

Fact: When rainforest areas are cut, biodiversity is destroyed also. Some rainforests have been evolving for tens of millions of years and contain species that cannot be found anywhere else. Destroying a rainforest eliminates many species—they become extinct.

Myth: Overpopulation is the major cause of tropical rainforest destruction, because poor farmers need to clear the land in order to grow food for survival.

Fact: "Blaming Third World peasants for rainforest destruction is like blaming foot soldiers for war," says the Rainforest Action Network. Unfair land distribution is the real culprit. Because wealthy landowners control most of the land, poor rural people have to clear rainforest land in order to farm. In Brazil, for example, 70 percent of the rural households are landless. Of all Brazilian landowners, 4.5 percent control 81 percent of the land.

Myth: If tropical rainforests are cleared, the land could be used to grow food crops, helping to eliminate starvation around the world.

Fact: The soil in tropical rainforests seldom can support monoculture (the production of one type of crop such as corn) for a long period of time unless there is extensive use of fertilizers and insecticides, which frequently creates dangerous pollution problems.

Myth: Clear-cutting—cutting a wide swath through a forest—is not much different from natural clearings created by fires and windfalls, and clear-cuts remove timber that can be used rather than left to rot.

Fact: In a natural clearing, fallen trees and other debris provide habitats for a great variety of species and supply nutrients for the growing vegetation, which is not the case in clear-cuts.

Myth: Deforestation is worse in South America than it is in North America.

Fact: Satellite photos contrasting deforestation in the Pacific Northwest and in Amazonia appear to show greater devastation in temperate rainforests.

precipitation in one temperate rainforest, diversity of species, and threats to temperate rainforests.

North American Rainforest Locations

The original extent of North American coastal temperate rainforests is shown in dark shading in Figure 1. According to estimates of some ecologists only 40 to 50 percent of these forests remain.

Precipitation

"On any given day in the Pacific coast rainforest, the odds are four to one that water will fall in one form or another"—as rain, mist,

FIGURE 1   Original Extent of Coastal Temperate Rainforests in North
America

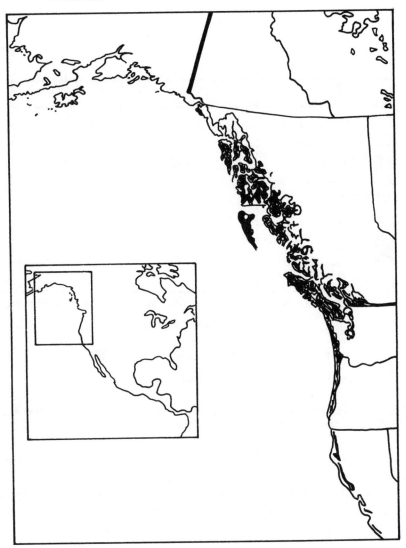

Source: Adapted by Douglas Gay from Ecotrust brochure, *The Rain Forests of Home*
(Portland, OR: Ecotrust, 1992).

fog, or snow. "On nearly every day of the year, the rainforest is
either absorbing moisture or releasing it," according to Karen
Kane, author of *America's Rain Forest*.

Kane also points out that an ancient tree in a temperate rainforest can support more than sixty million needles, which help
hold moisture. For example, moisture from fog collects on

FIGURE 2 Comparing Annual Rainfall in the Hoh Forest with Rainfall in Some U.S. Cities

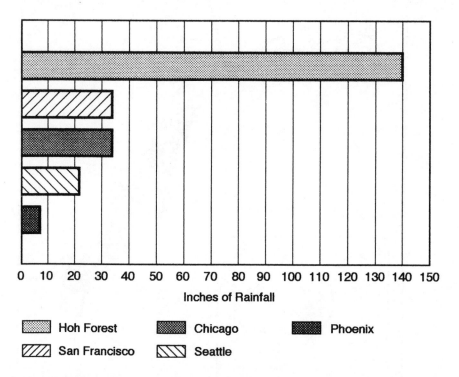

Inches of Rainfall

Hoh Forest   Chicago   Phoenix
San Francisco   Seattle

Source: Mary Lou Hanify and Craig Blencowe, Guide to the Hoh Rain Forest (Port Angeles, WA: Pacific Northwest National Parks and Forests Association, 1990), p. 8.

needles, dripping to the forest floor and accounting for 25 percent of the water supply in some areas. Figure 2 shows how the amount of annual rainfall in the Hoh Rainforest, part of the Olympic National Forest, contrasts with rainfall in selected U.S. cities.

Harvesting in U.S. Forests

In the United States, loggers use a variety of methods to harvest timber. Usually a silviculturalist, or timber manager—someone who manages a forest so that it will continue to yield timber— determines the harvesting system. A timber manager determines a harvest method by the type of ecosystem and the tree species in

TABLE 2   Harvesting and Logging Systems

Harvesting

Clear-cutting. Loggers cut all trees in an area ranging from 5 acres to several thousand acres; the U.S. Forest Service usually clear-cuts between 40 and 60 acres (16 and 24 hectares).
Shelterwood cutting. Loggers cut one-half to two-thirds of the trees in a stand, leaving the rest of the trees to shelter seedlings.
Seed tree cutting. Loggers cut nearly all trees on each acre, leaving just a few as a source of seeds but not enough to shelter seedlings.
Selection cutting. Loggers select a group of trees or individual trees in a stand for cutting.

Logging

Tractor logging. Large tractors drag logs down slopes and along trails to logging trucks that transport logs to mills. This is usually the most economical logging method but it often causes soil erosion.
High-lead logging. Lead cables are used to haul logs to a truck. Although more expensive than tractor logging, this method does less damage to soil.
Skyline logging. Heavy cables stretched between two points carry suspended logs to a landing area, preventing soil damage but increasing logging costs.
Helicopter and balloon logging. Logs suspended from a helicopter or balloon are airlifted to transport sites—the most expensive type of logging.

Source: "The Art of Silviculture," *The Citizens' Guide to Timber Management in the National Forests* (Eugene, OR: Cascade Holistic Economic Consultants, May 1989), pp. 12–21.

that system. A clear-cut method might be used, for example, if plenty of light is needed for natural regrowth of trees that do not reproduce well in heavy shade. The method also may be less costly than other harvesting systems, because after a clear-cut, seedlings planted by hand are apt to mature faster (and bring a quicker profit) than those allowed to grow back in a natural process. On the other hand, selecting a group of trees for harvest is not always the least expensive method, especially if roads have to be constructed into logging sites. Various silviculture systems and the methods used to remove trees or logs from national forests are described in Table 2.

Frequently, logging companies in the Pacific Northwest clear-cut large areas of forest, which can have a major impact on the exposed stands of forest on either side, as illustrated in Figure 3.

Many controversies have erupted over excessive logging in the temperate rainforests of the Pacific Northwest, and a variety of factors, such as bans on logging to protect endangered species and

FIGURE 3   Effects of Clear-Cutting

Clear-cutting in temperate forests allows sun and wind to penetrate the remaining stand, drying out and damaging root systems of trees.   During storms, unprotected trees on the fringe sometimes topple or are damaged.   Animals feed in clear-cut areas and frequently overgraze,   then enter the forest in search of food, which destroys part of the ecosystem. Illustration by   Brian D. Byrn.

less demand for timber, have led to decreases in timber harvests, as Figure 4 shows.

Wherever trees are cut in the United States, the timber is sawed, peeled, chipped, or burned to produce specific products. Table 3 shows what happens to cut timber.

The Value of Temperate Rainforests

One of the conflicts over logging in the Pacific Northwest has centered on the Pacific yew that grows in ancient forests. The yew's bark provides compounds that have been successfully used to treat malignant breast and ovarian tumors and may be effective in the treatment of other kinds of cancers.

Many varieties of the yew exist, and an expert on the subject, forester Hal Hartzell, Jr., of Oregon points out in his book *The Yew*

FIGURE 4   Timber Harvests in the Pacific Northwest

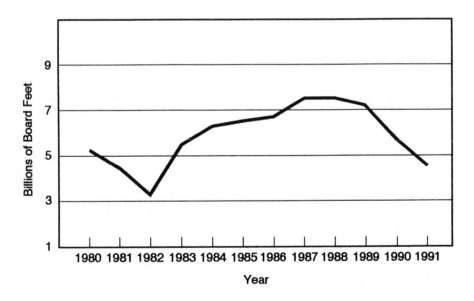

Source: U.S. Forest Service and Associated Press graphic "Harvests on National Forests," January 5, 1992.

TABLE 3   Processing Timber

· Saw the logs into smaller pieces and sell as lumber for construction or as raw material for manufacturing such products as doors and furniture.
· Peel the wood and glue veneers together to make plywood or to laminate onto other materials.
· Chip whole or partial logs to make particle board, hardboard, or wafer board, or to make pulp for the manufacture of paper products.
· Burn the wood for fuel.
· Use untreated logs for fence posts, supports in mines, and utility poles.

Source: "Product Mix," The Citizens' Guide to the Timber Industry and a Profile of U.S. Forests (Oak Grove, OR: Cascade Holistic Economic Consultants, July 1991), pp. 22–24.

TABLE 4   Yews Providing Anticancer Compounds

(With Their Common Names or Descriptions)
T. baccata (European yew)
    T. b. barronii (its leaves change to coppery color with age)
    T. b. fastigiata (Irish yew)
    T. b. repandens (a small spreading bush with flat sickle-shaped leaves)
T. brevifolia (Pacific yew)
T. canadensis (Canadian yew)
T. chinensis (Chinese yew)
T. cuspidata(Japanese yew)
    T. c. capitata(upright form with green leaves)
    T. c. nana (a small wide-spreading shrub)
T. floridana (Florida yew)
T. media (a hybrid with olive green branches)
    T. m. densiformis (a dense shrub with bright green leaves)
    T. m. hicksii (a columnar shaped bush)
T. sumatrana (Southeast Asian Yew)
T. wallichiana (Himalayan Yew)

Source: Hal Hartzell, Jr., *The Yew Tree* (Eugene, OR: Hulogosi, 1991), pp. 293–298.

Tree—A Thousand Whispers that the medicinal benefits of the yew have been known for centuries. "North American natives used yew as an abortifacient and [to combat] illnesses such as bronchitis, scurvy and skin cancer," Hartzell explained. He has compiled a list of yew species and cultivars that contain taxol or convertible taxanes, ingredients in anticancer drugs. Some of those trees are described in Table 4.

Along with the Pacific yew, many other types of evergreens such as the Douglas fir, Sitka Spruce, and Western Hemlock (Washington's state tree) and broadleaf trees such as the Red Adler, sometimes mistaken for birch, and Bigleaf Maple grow in the ancient forests of the Pacific Northwest. But the ancient forests

---

TABLE 5   Destruction of Ancient Forests: What Will Be Lost?

· Water for millions of people. Forests help maintain clean, abundant water supplies for tens of millions of people and for agricultural and industrial purposes.
· Spawning grounds for salmon. Rivers and streams in the forest provide spawning grounds for salmon fisheries and habitat for trout and other sport fish.
· Recreation. Every year, millions of people visit the Pacific Northwest for camping, hiking, fishing, touring, and similar activities, and destruction of the forests threatens a recreational industry worth several billion dollars per year.
· Abundant wildlife. Ancient forest ecosystems support more than 200 animal and innumerable plant species. Destroying the forests could destroy species of plants and animals forever.
· Lifesaving medicinals. As described previously, ancient forests are habitats for the Pacific yew and perhaps still-to-be-discovered plants that provide chemicals for medicinal purposes, which will never be known if the forests are destroyed.

Adapted from Western Ancient Forest Campaign, *America's Ancient Forests* (undated brochure).

TABLE 6   Life from a "Dead" Log

✓ On wet ground, a fallen log holds moisture and prevents soil erosion.
✓ Plants and tree seedlings take root in moist, fertile soil near a fallen log or within the log itself.
✓ Spiders use cracks in a log for nests.
✓ Beetles bore into a log, making a way for other organisms to spread through the rest of the wood.
✓ Termites build a maze of tunnels and chambers in the log.
✓ Various scavengers feed on loose bark.
✓ In a stream, a fallen log creates protected pools where fish spawn and rapids where insects thrive and serve as food for fish.
✓ As a log rots, it becomes spongelike, holding moisture and helping to provide nutrients for soil and plants.

---

are much more than trees. A great variety of plant life thrives in temperate rainforests, as do diverse species of mammals and birds. Most of the books on temperate rainforests annotated in Chapter 6 contain detailed descriptions of plant and animal life found in ancient forests. If those ecosystems are destroyed, Americans stand to lose a great deal, as Table 5 illustrates.

After hundreds of years of life, a Douglas fir tree in an ancient forest dies of old age and topples over. But the log that rests on the forest floor or drops into a stream continues to provide life, acting as a host for countless living creatures and serving many other functions, as illustrated in Table 6.

Tropical Rainforests

Although many people have a general understanding of what a tropical rainforest is, their concepts vary, particularly in terms of the value of rainforests, as is shown in the diagrams and charts in this section. The general location of tropical rainforests is fairly well known also, but the map of the world's rainforests included in this section graphically shows the small percentage of the world's land area that these forests cover, and other maps and charts depict how forest cover has vanished.

Diverse Images of a Tropical Rainforest

A tropical rainforest is made up of several layers that form distinct habitats known from the top down as the "Emergent Trees," "Upper Canopy," "Lower Canopy," and the "Understory." Figure 5 depicts examples of the varied life in these separate habitats.

Whether people want to exploit or conserve tropical rainforest resources, they have diverse views on how a rainforest should be used. Figure 6 illustrates some of those views.

Although people view rainforests in diverse ways, a global perspective on tropical rainforests is emerging, according to the editors of *Lessons of the Rainforest*. The editors summarized components of that worldview, and extracts appear in Table 7.

Where Disappearing Rainforests Are Located

Whatever the views about preserving rainforests, the fact is they are disappearing. Tropical rainforests cover only a small portion of the world's land area, as Figure 7 shows. Figure 8 shows Costa Rica's loss of forest (primarily rainforest) cover during a 44-year period.

In the Amazonia rainforest, the highest level of habitat is marked by the emergent trees—131 feet or more above the ground—which is a critical site for the Guiana crested eagle and the harpy eagle (both endangered) when they search for prey. Other birds help scatter plant seeds in their droppings as they fly over the emergent trees and the upper canopy, a nearly unbroken layer of vegetation from about 66 feet to 82 feet high that captures sunlight and carries out the process of photosynthesis. Many animals, including howler monkeys, fruit bats, and sloths, and birds such as parrots spend much of their lives here. The lower canopy is the habitat for many animals making their way to higher levels and birds, butterflies, spiders, and iguanas forage along the branches. In the understory, the lowest level of the rainforest, are various turtles and other reptiles, frogs, fishing bats, hummingbirds, and large mammals such as deer. Illustration by Brian D. Byrn. ▶

FIGURE 5   The Habitats of a Rainforest

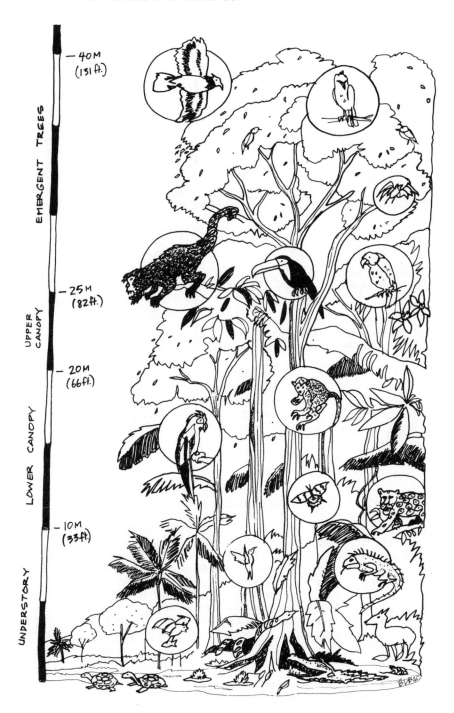

FIGURE 6   Uses for a Tropical Rainforest

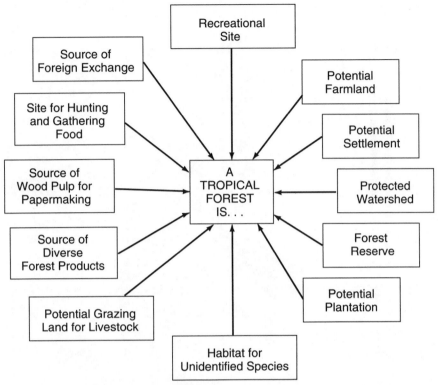

Source: Adapted from Kenneth A. Longman and Jan Jenik, *Tropical Forest and Its Environment*, (White Plains, NY: Longman Publishing Group, 1987), p. 8.

TABLE 7   A Global View of Rainforests

1. The rainforests are fragile, nonrenewable resources.
2. The rainforests are poorly understood by the Western world.
3. The forests are home to indigenous peoples and other forest dwellers.
4. Human welfare and forest conservation are inextricably linked.
5. Tropical rainforest development is plagued with problems.
6. Rainforests occur primarily in countries with developing economies.
7. There are many alternatives to forest clearing.
8. Tropical deforestation is a global problem requiring global solutions.
9. The rainforest issue is a hologram that presents an urgent and powerful challenge to the dominant worldview.
10. Future generations will find it incomprehensible if we do not act.

Source: Suzanne Head and Robert Heinzman, eds., *Lessons of the Rainforest* (San Francisco: Sierra Club Books, 1990), pp. 7–9.

FIGURE 7 Tropical Rainforests of the World

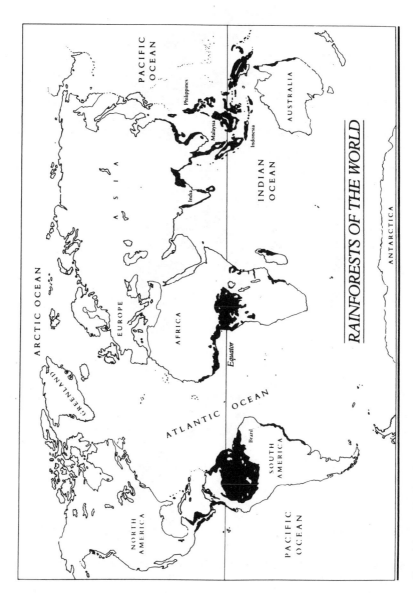

RAINFORESTS OF THE WORLD

FIGURE 8   Diminishing Forest Cover in Costa Rica, 1940–1984

Source: Adapted by Douglas Gay from Donella H. Meadows, Dennis L. Meadows, and Jorgen Randers, *Beyond the Limits* (Post Mills, VT: Chelsea Green, 1992), p. 59.

FIGURE 9   Tropical Deforestation

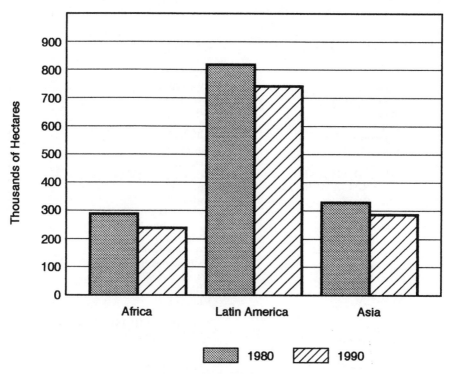

Source: FAO of the United Nations and "Global Frontiers: A Reprint," *The Christian Science Monitor* (special reprint of an article series, April–May 1991), p. 5.

Figure 9 illustrates changes in tropical deforestation from 1980 to 1990 in Africa, Latin America, and Asia. Tables 8 and 9 depict where deforestation has taken place.

Norman Myers, a British ecologist, has identified 18 tropical regions, called "hot spots," that need special protection because they are rich habitats for rapidly disappearing plant and animal species. Table 9 shows twelve of the hot spots and the percentage of standing forest in each.

According to Brazilian government sources, the rate of deforestation in that country has been dropping steadily since 1985, as Figure 10 shows, but it is not clear whether the trend will continue.

The Value of Tropical Rainforests

One of the most compelling reasons for preserving tropical rainforests is the great many benefits forests provide. Along with

TABLE 8 Status of Tropical Rainforests

Key: S = some forests
      < = vanishing forests
      0 = forests almost gone

SOUTH AMERICA
Boliva S
Brazil <
Colombia 0
French Guiana S
Guyanas S
Peru S
Surinam S
Venezuela <

CARIBBEAN AND CENTRAL AMERICA
Belize S
Costa Rica S
Guatemala <
Honduras S
Mexico 0
Nicaragua S
Panama <

AFRICA
Angola S
Benin S
Cameroon 0
Central African Republic S
Equatorial Guinea S
Gabon <
Ghana 0
Ivory Coast 0

Liberia 0
Madagascar 0
Nigeria 0
Rwanda S
Sierra Leone 0
Zaire S

ASIA
Bangladesh <
Brunei S
Burma S
China <
India S
Indonesia 0
Kampuchea S
Malaysia 0
Papua New Guinea S
Philippines 0
Sri Lanka <
Thailand <
Vietnam 0

THE PACIFIC ISLANDS AND AUSTRALIA
Hawaii 0
Queensland 0
Tasmania <
Fiji <
Solomon Islands S

Source: John Nichol, *The Mighty Rain Forest* (London: David & Charles, 1990), pp. 138–139.

TABLE 9 Tropical Forest "Hot Spots"

"Hot Spots"	Percentage of Forest Standing
Central America	18
Colombian Choco	72
Western Ecuador	4
Tropical Andes	35
Atlantic Forest (Brazil)	2
Upper Guinean Forest	3
Eastern Arc Mts.(Tanzania)	19
Eastern Himalayas	16
Malaysia	29
Madagascar	16
Philippines	3
Indonesia	43

Source: "Hot Spots under Intense Heat (diagram)," *National Wildlife* (April–May 1992), p. 45.

FIGURE 10   Deforestation in Brazil

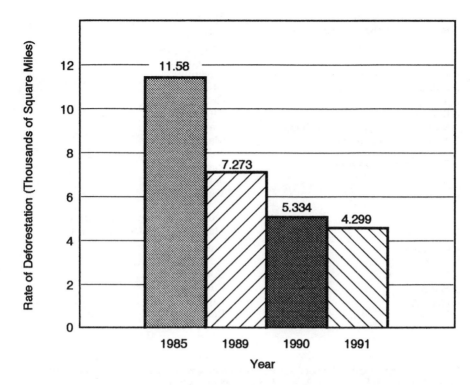

Source: Brazil's National Institute of Space Research and ˙*The New York Times* (May 22, 1992), p. A4.

TABLE 10 Examples of Tropical Rainforest Products

Fibers, Gums, and Resins (and their uses)

Bamboo (furniture)
Jute (rope)
Kapok (insulation)
Raffia (baskets)
Ramie (fabric)
Rattan (furniture)
Rubber latex (rubber products)
Tung oil (wood finisher)

Foods and Beverages

Avocado
Banana
Brazil nuts
Cashew nuts
Coffee
Grapefruit
Guava
Okra
Papaya
Peanuts
Peppers
Pineapple
Tea

Spices

Allspice
Black Pepper
Cardamon
Cayenne
Chili
Cinnamon
Cloves
Ginger
Mace
Nutmeg
Paprika
Sesame seeds
Vanilla

Oils for Cosmetics, Perfumes, Flavorings, and Cough Drops

Camphor
Cascarilla
Coconut
Eucalyptus
Palm
Rosewood

Sources: "Our Tropical Connection," *Environmental Almanac* (Boston: Houghton Mifflin, 1991), p. 284 and Rainforest Action Network (undated brochure).

tropical woods and medicinals from rainforest plants, a variety of products, such as oils and resins, now used in industry and those that will be developed for industrial purposes in the future depend on rainforest sources. Hundreds of other products that people use in their daily lives also come from tropical rainforests, as Table 10 shows. Some fruits, vegetables, and fibers originated in rainforests but through crossbreeding have become part of agricultural production in industrialized nations. In fact, crop breeders need genes from wild plants and primitive crops to develop hardier modern varieties—those that can resist disease or other threats.

The use of tropical wood for commercial purposes is one factor contributing to tropical forest loss, and the amount of tropical wood cut for commercial use is expected to increase, as shown in Figure 11. Such tropical woods as balsa, mahogany, rosewood,

FIGURE 11   Tropical Wood Cut for Commercial Use

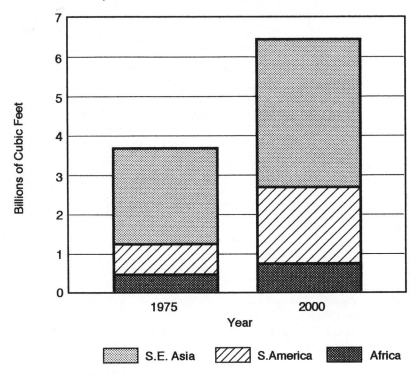

Source: American Forestry Association map, 1988.

sandalwood, and teak are used to make furniture, cabinets, doors, flooring, packing cases, boats, drawing boards, and toys. Some countries that have exported tropical hardwoods for years are now importing them, and the Philippines, Malaysia, the Ivory Coast, Nigeria, and Thailand are expected to exhaust their supplies of timber for export by 2000.

In November 1991, the environmental organization Friends of the Earth (FoE) launched a campaign in the United Kingdom to persuade businesses to stop selling timber products that contribute to the destruction of rainforests. The UK is one of the main consumers of timber products, and FoE urged British consumers to boycott the purchase of these goods until stores take positive steps to stop selling them.

To follow up on the boycott campaign, 60 local FoE groups conducted a nationwide survey to determine consumer attitudes about buying tropical wood products. A total of 15,154 consumers were included in the survey but the number responding to each

---

TABLE 11   Friends of the Earth British Survey Questions and Responses

A. Are you concerned about the destruction of tropical rainforests? (No. of answers: 15,154)

Yes: 91%	No: 5.5%	Don't know: 3.5%
(13,838)	(806)	(510)

B. Do you ever buy wood, or wooden furniture or fittings? (No. of answers 15,019)

Yes: 64%	No: 34%	Don't know: 2%
(9593)	(5153)	(273)

If answer is "No", respondents proceed to question E.

C. Would you be able to tell if you were buying products made from rainforest timber? (Those who answered "No" to question B were excluded; No. of answers 10,272)

Yes: 33%	No: 60%	Don't know: 7%
(3384)	(6126)	(762)

D. If you knew that a timber product came from rainforests, would you still buy it? (Those who answered "No" to question B were excluded; No. of answers 10,277)

Yes: 22%	No: 58%	Don't know: 20%
(2290)	(6001)	(1986)

E. Do you think that shops should label rainforest timber products? (No. of answers 14,663)

Yes: 93%	No: 4%	Don't know: 3%
(13,581)	(632)	(450)

F. Do you think that shops should sell timber products which contribute to rainforest destruction when a viable alternative exists? (No. of answers 15,076)

Yes: 10%	No: 83%	Don't know: 7%
(1541)	(12,459)	(1076)

---

question varied. The results from the survey, as shown in Table 11, were published in March 1992 and showed that over 90 percent of the public believed businesses should face up to their environmental responsibilities and stop selling tropical timber products that contribute to rainforest destruction. More than half of the respondents said they would not buy wood products if they knew the timber came from rainforests, but few products carry labels showing the source of raw materials.

FIGURE 12   Where Is Belize?

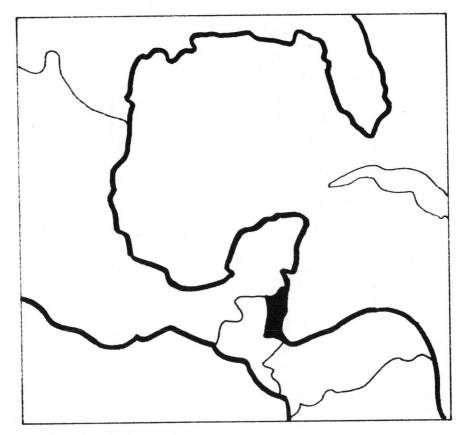

Source: Map adapted by Douglas Gay from Belizean Rain Forest Products brochure (undated).

Belize, formerly known as British Honduras, became an independent country in 1981. Shown in Figure 12, Belize is about the size of New Hampshire and contains 16 national rainforest reserves.

In 1992, after more than 30 years of campaigning by environmental, anthropological, legal, and human rights groups, the Brazilian government set aside 68,331 square miles (176,977 square kilometers) of Amazon wilderness for the Yanomami people. The Brazilian reserve is adjacent to a similar reserve for the indigenous people in Venezuela. Figure 13 shows the location of the reserve in South America.

FIGURE 13   The Yanomami Reserve in South America

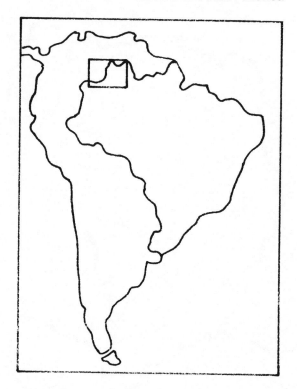

Source: Map adapted by Douglas Gay from Survival International's campaign brochure (undated).

Atmospheric Effects of Carbon Dioxide Emissions

The atmosphere consists of several gases, primarily water vapor and small amounts of carbon dioxide (CO_2) plus several other gases. Short wave-length radiation from the sun can penetrate the gases to reach the Earth and warm the surface. But the heat that radiates back from the Earth is long wave-length radiation, which is absorbed by the gases and returned to the Earth. When gases build up in the atmosphere, they trap more and more heat and create what is known as the greenhouse effect, which scientists theorize leads to a rise in global temperatures.

Although only small amounts of CO_2 make up the atmosphere, it is the major component of the so-called greenhouse

FIGURE 14   Rising Levels of Carbon Dioxide Concentration

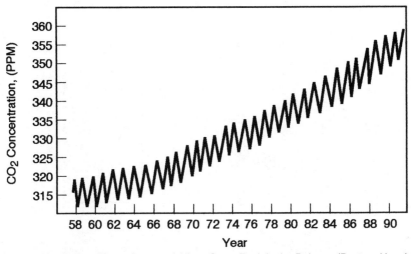

Sources: Mauna Loa Observatory and Albert Gore, *Earth in the Balance* (Boston: Houghton Mifflin, 1992), p. 5.

gases. The concentrations of CO_2 have been steadily increasing since they were first measured in 1958 at the Mauna Loa Observatory in Hawaii. Figure 14 shows the monthly average of CO_2 concentrations from 1958 to mid-1991. Concentrations dip in the summer months because vegetation absorbs large amounts of CO_2, but in the winter months when there is less green vegetation, the gases build up again.

Global deforestation accounts for a great portion of the buildup of CO_2 in the atmosphere, but the three major industrial nations—the United States, Japan, and Germany—release large amounts of greenhouse gases from manufacturing facilities, transportation, and other human activities. Since these nations account for only 8.5 percent of the world's population (see Figures 15 and 16), they are considered prime polluters, contributing disproportionately to the greenhouse effect.

Documents on Rainforests

A variety of documents and resolutions have been published since the 1980s to help focus public attention on rainforest issues and to

FIGURE 15   Three Major Industrial Nations and Their CO_2 Emissions

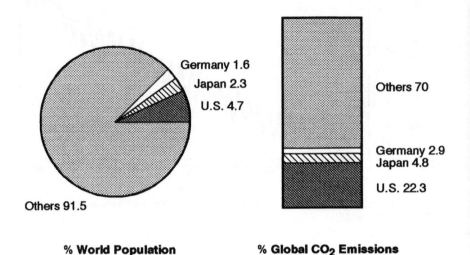

Germany 1.6
Japan 2.3
U.S. 4.7

Others 70

Germany 2.9
Japan 4.8

U.S. 22.3

Others 91.5

% World Population   **% Global CO$_2$ Emissions**

Sources: World Resources Institute and *The New York Times* (May 31, 1992), p. 6E.

call for action to stop rainforest devastation. While these publications are not legal documents, they do help illuminate the many problems associated with the destruction of rainforests. One example is "An Emergency Call to Action for the Forests and Their People" issued in 1990 by the World Rainforest Movement in Malaysia. It is a declaration calling on the United Nations and national governments to implement specific measures such as these:

1. To empower forest peoples and those who depend upon the forests for their livelihood with the responsibility for safeguarding the forests and ensuring their regeneration . . . ;
2. To halt all those practices and projects which would contribute either directly or indirectly to further forest loss;
3. To revise radically the policies of those agencies that currently finance the projects and practices causing deforestation . . . ;
4. To implement, through the agency of forest peoples and under their direction, a programme for regenerating degraded forest lands and reinvigorating local cultures;
5. To take immediate steps to curb the wastage, misuse and overconsumption of timber products;

FIGURE 16   Carbon Dioxide Emissions: Selected Nations

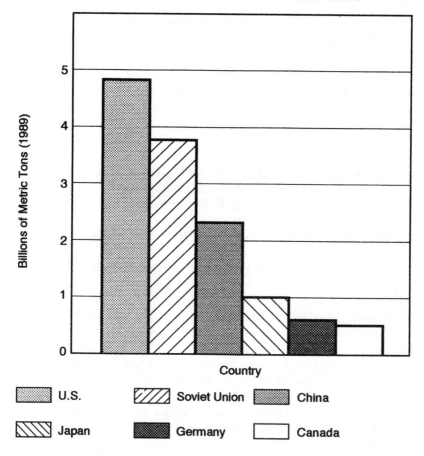

Sources: World Resources Institute and *The Oregonian* (June 1, 1992), p. A3.

6. To ban all imports of tropical timber and tropical wood products from natural forests;

7. To take immediate steps to cut down the consumption of imported beef from tropical forest areas;

8. To take immediate steps to reduce atmospheric air pollution and to eliminate those pollutants responsible for forest [destruction] . . . ;

9. to restructure the present unequal world economic system . . . that favour[s] the developed countries at the expense of the poor . . . ;

10. To initiate a global shift towards developing sustainable livelihoods. . . .

Source: World Rainforest Movement, *Rainforest Destruction* (Penang, Malaysia: World Rainforest Movement, 1992), pp. 8–14.

United Nations Documents

Two of the five major documents that were signed by heads of state and were the result of the Earth Summit held in Rio de Janeiro in June 1992 were the "Rio Declaration" and a statement of forest principles. The Rio Declaration presents basic guidelines for nations to follow in matters of environmental protection and development; the forest principles are basic guidelines for the protection of all types of forests.

REPORT OF THE UNITED NATIONS CONFERENCE ON ENVIRONMENT AND DEVELOPMENT
(Rio de Janeiro, 3–14 June 1992)
Annex I
RIO DECLARATION ON ENVIRONMENT AND DEVELOPMENT

The United Nations Conference on Environment and Development,

Having met at Rio de Janeiro from 3 to 14 June 1992,

Reaffirming the Declaration of the United Nations Conference on the Human Environment, adopted at Stockholm on 16 June 1972, and seeking to build upon it,

With the goal of establishing a new and equitable global partnership through the creation of new levels of cooperation among States, key sectors of societies and people,

Working towards international agreements which respect the interests of all and protect the integrity of the global environmental and developmental system,

Recognizing the integral and interdependent nature of the Earth, our home,

Proclaims that:

Principle 1

Human beings are at the centre of concerns for sustainable development. They are entitled to a healthy and productive life in harmony with nature.

Principle 2

States have, in accordance with the Charter of the United Nations and the principles of international law, the sovereign right to exploit their own resources pursuant to their own environmental and developmental policies, and the responsibility to ensure that activities within their jurisdiction or control do not cause damage to the environment of other States or of areas beyond the limits of national jurisdiction.

Principle 3

The right to development must be fulfilled so as to equitably meet developmental and environmental needs of present and future generations.

Principle 4

In order to achieve sustainable development, environmental protection shall constitute an integral part of the development process and cannot be considered in isolation from it.

Principle 5

All States and all people shall cooperate in the essential task of eradicating poverty as an indispensable requirement for sustainable development, in order to decrease the disparities in standards of living and better meet the needs of the majority of the people of the world.

Principle 6

The special situation and needs of developing countries, particularly the least developed and those most environmentally vulnerable, shall be given special priority. International actions in the field of environment and development should also address the interests and needs of all countries.

Principle 7

States shall cooperate in a spirit of global partnership to conserve, protect and restore the health and integrity of the Earth's ecosystem. In view of the different contributions to global environmental degradation, States have common but differentiated responsibilities. The developed countries acknowledge the responsibility that they bear in the international pursuit of sustainable development in view of the pressures their societies place on the global environment and of the technologies and financial resources they command.

Principle 8

To achieve sustainable development and a higher quality of life for all people, States should reduce and eliminate unsustainable patterns of production and consumption and promote appropriate demographic policies.

Principle 9

States should cooperate to strengthen endogenous capacity-building for sustainable development by improving scientific understanding through exchanges of scientific and technological knowledge, and by enhancing the development, adaptation, diffusion and transfer of technologies, including new and innovative technologies.

Principle 10

Environmental issues are best handled with the participation of all concerned citizens, at the relevant level. At the national level, each individual shall have appropriate access to information concerning the environment that is held by public authorities, including information on hazardous materials and activities in their communities, and the opportunity to participate in decision-making processes. States shall facilitate and encourage public awareness and participation by making information widely available. Effective access to judicial and administrative proceedings, including redress and remedy, shall be provided.

Principle 11

States shall enact effective environmental legislation. Environmental standards, management objectives and priorities should reflect the environmental and developmental context to which they apply. Standards applied by some countries may be inappropriate and of unwarranted economic and social cost to other countries, in particular developing countries.

Principle 12

States should cooperate to promote a supportive and open international economic system that would lead to economic growth and sustainable development in all countries, to better address the problems of environmental degradation. Trade policy measures for environmental purposes should not constitute a means of arbitrary or unjustifiable discrimination or a disguised restriction on international trade. Unilateral actions to deal with environmental challenges outside the jurisdiction of the importing country should be avoided. Environmental measures addressing transboundary or global environmental problems should, as far as possible, be based on an international consensus.

Principle 13

States shall develop national law regarding liability and compensation for the victims of pollution and other environmental damage. States shall also cooperate in an expeditious and more determined manner to develop further international law regarding liability and compensation for adverse effects of environmental damage caused by activities within their jurisdiction or control to areas beyond their jurisdiction.

Principle 14

States should effectively cooperate to discourage or prevent the relocation and transfer to other States of any activities and substances that cause severe environmental degradation or are found to be harmful to human health.

Principle 15

In order to protect the environment, the precautionary approach shall be widely applied by States according to their capabilities. Where there are threats of serious or irreversible damage, lack of full scientific certainty shall not be used as a reason for postponing cost-effective measures to prevent environmental degradation.

Principle 16

National authorities should endeavour to promote the internalization of environmental costs and the use of economic instruments, taking into account the approach that the polluter should, in principle, bear the cost of pollution, with due regard to the public interest and without distorting international trade and investment.

Principle 17

Environmental impact assessment, as a national instrument, shall be undertaken for proposed activities that are likely to have a significant adverse impact on the environment and are subject to a decision of a competent national authority.

Principle 18

States shall immediately notify other States of any natural disasters or other emergencies that are likely to produce sudden harmful effects on the environment of those States. Every effort shall be made by the international community to help States so afflicted.

Principle 19

States shall provide prior and timely notification and relevant information to potentially affected States on activities that may have a significant adverse transboundary environmental effect and shall consult with those States at an early stage and in good faith.

Principle 20

Women have a vital role in environmental management and develoment. Their full participation is therefore essential to achieve sustainable development.

Principle 21

The creativity, ideals and courage of the youth of the world should be mobilized to forge a global partnership in order to achieve sustainable development and ensure a better future for all.

Principle 22

Indigenous people and their communities and other local communities have a vital role in environmental management and development because of their knowledge and traditional practices. States should recognize and duly support their identity, culture and interests and enable their effective participation in the achievement of sustainable development.

Principle 23

The environment and natural resources of people under oppression, domination and occupation shall be protected.

Principle 24

Warfare is inherently destructive of sustainable development. States shall therefore respect international law providing protection for the environment in times of armed conflict and cooperate in its further development, as necessary.

Principle 25

Peace, development and environmental protection are inter-dependent and indivisible.

Principle 26

States shall resolve all their environmental disputes peacefully and by appropriate means in accordance with the Charter of the United Nations.

Principle 27

States and people shall cooperate in good faith and in a spirit of partnership in the fulfillment of the principles embodied in this Declaration and in the further development of international law in the field of sustainable development.

REPORT OF THE UNITED NATIONS CONFERENCE ON ENVIRONMENT AND DEVELOPMENT
(Rio de Janeiro, 3–14 June 1992)
Annex III
NON-LEGALLY BINDING AUTHORITATIVE STATEMENT OF PRINCIPLES FOR A GLOBAL CONSENSUS ON THE MANAGEMENT, CONSERVATION AND SUSTAINABLE DEVELOPMENT OF ALL TYPES OF FORESTS

Preamble

(a) The subject of forests is related to the entire range of environmental and development issues and opportunities, including the right to socio-economic development on a sustainable basis.

(b) The guiding objective of these principles is to contribute to the management, conservation and sustainable development of forests and to provide for their multiple and complementary functions and uses.

(c) Forestry issues and opportunities should be examined in a holistic and balanced manner within the overall context of environment and development, taking into consideration the multiple functions and uses of forests, including traditional uses, and the likely economic and social stress when these uses are constrained or restricted, as well as the potential for development that sustainable forest management can offer.

(d) These principles reflect a first global consensus on forests. In committing themselves to the prompt implementation of these principles, countries also decide to keep them under assessment for their adequacy with regard to further international cooperation on forest issues.

(e) These principles should apply to all types of forests, both natural and planted, in all geographical regions and climatic zones, including austral, boreal, subtemperate, temperate, subtropical and tropical.

(f) All types of forests embody complex and unique ecological processes which are the basis for their present and potential capacity to provide resources to satisfy human needs as well as environmental values, and as such their sound management and conservation is of concern to the Governments of the countries to which they belong and are of value to local communities and to the environment as a whole.

(g) Forests are essential to economic development and the maintenance of all forms of life.

(h) Recognizing that the responsibility for forest management, conservation and sustainable development is in many States allocated among federal/national, state/provincial and local levels of government, each State, in accordance with its constitution and/or national legislation, should pursue these principles at the appropriate level of government.

Principles/Elements

1. (a) States have, in accordance with the Charter of the United Nations and the principles of international law, the sovereign right to exploit their own resources pursuant to their own environmental policies and have the responsibility to ensure that activities within their jurisdiction or control do not cause damage to the environment of other States or of areas beyond the limits of national jurisdiction.

   (b) The agreed full incremental cost of achieving benefits associated with forest conservation and sustainable development requires increased international cooperation and should be equitably shared by the international community.

2. (a) States have the sovereign and inalienable right to utilize, manage and develop their forests in accordance with their development needs and level of socio-economic development and on the basis of national policies consistent with sustainable development and legislation, including the conversion of such areas for other uses within the overall socio-economic development plan and based on rational land-use policies.

   (b) Forest resources and forest lands should be sustainably managed to meet the social, economic, ecological, cultural and spiritual needs of present and future generations. These needs are for forest products and services, such as wood and wood products, water, food, fodder, medicine, fuel, shelter, employment, recreation, habitats for wildlife, landscape diversity, carbon sinks and reservoirs, and for other forest products. Appropriate measures should be taken to protect forests against harmful effects of pollution, including air-borne pollution, fires, pests and diseases, in order to maintain their full multiple value.

   (c) The provision of timely, reliable and accurate information on forests and forest ecosystems is essential for public understanding and informed decision-making and should be ensured.

   (d) Governments should promote and provide opportunities for the participation of interested parties, including local communities and indigenous people, industries, labour, non-governmental organizations and individuals, forest dwellers and women, in the development, implementation and planning of national forest policies.

3. (a) National policies and strategies should provide a framework for increased efforts, including the development and strengthening of institutions and programmes for the management, conservation and sustainable development of forests and forest lands.

 (b) International institutional arrangements, building on those organizations and mechanisms already in existence, as appropriate, should facilitate international cooperation in the field of forests.

 (c) All aspects of environmental protection and social and economic development as they relate to forests and forest lands should be integrated and comprehensive.

4. The vital role of all types of forests in maintaining the ecological processes and balance at the local, national, regional and global levels through, inter alia, their role in protecting fragile ecosystems, watersheds and freshwater resources and as rich storehouses of biodiversity and biological resources and sources of genetic material for biotechnology products, as well as photosynthesis, should be recognized.

5. (a) National forest policies should recognize and duly support the identity, culture and the rights of indigenous people, their communities and other communities and forest dwellers. Appropriate conditions should be promoted for these groups to enable them to have an economic stake in forest use, perform economic activities, and achieve and maintain cultural identity and social organization, as well as adequate levels of livelihood and well-being, through, inter alia, those land tenure arrangements which serve as incentives for the sustainable management of forests.

 (b) The full participation of women in all aspects of the management, conservation and sustainable development of forests should be actively promoted.

6. (a) All types of forests play an important role in meeting energy requirements through the provision of a renewable source of bio-energy, particularly in developing countries, and the demands for fuelwood for household and industrial needs should be met through sustainable forest management, afforestation and reforestation. To this end, the potential contribution of plantations of both indigenous and introduced species for the provision of both fuel and industrial wood should be recognized.

 (b) National policies and programmes should take into account the relationship, where it exists, between the conservation, management and sustainable development of forests and all aspects related to the production, consumption, recycling and/or final disposal of forest products.

(c) Decisions taken on the management, conservation and sustainable development of forest resources should benefit, to the extent practicable, from a comprehensive assessment of economic and non-economic values of forest goods and services and of the environmental costs and benefits. The development and improvement of methodologies for such evaluations should be promoted.

(d) The role of planted forests and permanent agricultural crops as sustainable and environmentally sound sources of renewable energy and industrial raw material should be recognized, enhanced and promoted. Their contribution to the maintenance of ecological processes, to offsetting pressure on primary/old-growth forest and to providing regional employment and development with the adequate involvement of local inhabitants should be recognized and enhanced.

(e) Natural forests also constitute a source of goods and services, and their conservation, sustainable management and use should be promoted.

7. (a) Efforts should be made to promote a supportive international economic climate conducive to sustained and environmentally sound development of forests in all countries, which include, inter alia, the promotion of sustainable patterns of production and consumption, the eradication of poverty and the promotion of food security.

(b) Specific financial resources should be provided to developing countries with significant forest areas which establish programmes for the conservation of forests including protected natural forest areas. These resources should be directed notably to economic sectors which would stimulate economic and social substitution activities.

8. (a) Efforts should be undertaken towards the greening of the world. All countries, notably developed countries, should take positive and transparent action towards reforestation, afforestation and forest conservation, as appropriate.

(b) Efforts to maintain and increase forest cover and forest productivity should be undertaken in ecologically, economically and socially sound ways through the rehabilitation, reforestation and re-establishment of trees and forests on unproductive, degraded and deforested lands, as well as through the management of existing forest resources.

(c) The implementation of national policies and programmes aimed at forest management, conservation and sustainable development, particularly in developing countries, should be supported by international financial and technical cooperation, including through the private sector, where appropriate.

(d) Sustainable forest management and use should be carried out in accordance with national development policies and priorities and on the basis of environmentally sound national guidelines. In the formulation of such guidelines, account should be taken, as appropriate and if applicable, of relevant internationally agreed methodologies and criteria.

(e) Forest management should be integrated with management of adjacent areas so as to maintain ecological balance and sustainable productivity.

(f) National policies and/or legislation aimed at management, conservation and sustainable development of forests should include the protection of ecologically viable representative or unique examples of forests, including primary/old-growth forests, cultural, spiritual, historical, religious and other unique and valued forests of national importance.

(g) Access to biological resources, including genetic material, shall be with due regard to the sovereign rights of the countries where the forests are located and to the sharing on mutually agreed terms of technology and profits from biotechnology products that are derived from these resources.

(h) National policies should ensure that environmental impact assessments should be carried out where actions are likely to have significant adverse impacts on important forest resources, and where such actions are subject to a decision of a competent national authority.

9. (a) The efforts of developing countries to strengthen the management, conservation and sustainable development of their forest resources should be supported by the international community, taking into account the importance of redressing external indebtedness, particularly where aggravated by the net transfer of resources to developed countries, as well as the problem of achieving at least the replacement value of forests through improved market access for forest products, especially processed products. In this respect, special attention should also be given to the countries undergoing the process of transition to market economies.

(b) The problems that hinder efforts to attain the conservation and sustainable use of forest resources and that stem from the lack of alternative options available to local communities, in particular the urban poor and poor rural populations who are economically and socially dependent on forests and forest resources, should be addressed by Governments and the international community.

(c) National policy formulation with respect to all types of forests should take account of the pressures and demands imposed on forest ecosystems and resources from influencing factors outside the forest sector, and intersectoral means of dealing with these pressures and demands should be sought.

10. New and additional financial resources should be provided to developing countries to enable them to sustainably manage, conserve and develop their forest resources, including through afforestation, reforestation and combating deforestation and forest and land degradation.

11. In order to enable, in particular, developing countries to enhance their endogenous capacity and to better manage, conserve and develop their forest resources, the access to and transfer of environmentally sound technologies and corresponding know-how on favourable terms, including on concessional and preferential terms, as mutually agreed, in accordance with the relevant provisions of Agenda 21, should be promoted, facilitated and financed, as appropriate.

12. (a) Scientific research, forest inventories and assessments carried out by national institutions which take into account, where relevant, biological, physical, social and economic variables, as well as technological development and its application in the field of sustainable forest management, conservation and development, should be strengthened through effective modalities, including international cooperation. In this context, attention should also be given to research and development of sustainably harvested non-wood products.

   (b) National and, where appropriate, regional and international institutional capabilities in education, training, science, technology, economics, anthropology and social aspects of forests and forest management are essential to the conservation and sustainable development of forests and should be strengthened.

   (c) International exchange of information on the results of forest and forest management research and development should be enhanced and broadened, as appropriate, making full use of education and training institutions, including those in the private sector.

   (d) Appropriate indigenous capacity and local knowledge regarding the conservation and sustainable development of forests should, through institutional and financial support and in collaboration with the people in the local communities concerned, be recognized, respected, recorded, developed and, as appropriate, introduced in the implementation of programmes. Benefits arising from the utilization of indigenous knowledge should therefore be equitably shared with such people.

13. (a) Trade in forest products should be based on non-discriminatory and multilaterally agreed rules and procedures consistent with international trade law and practices. In this context, open and free international trade in forest products should be facilitated.

    (b) Reduction or removal of tariff barriers and impediments to the provision of better market access and better prices for higher value-added forest products and their local processing should be encouraged to enable producer countries to better conserve and manage their renewable forest resources.

    (c) Incorporation of environmental costs and benefits into market forces and mechanisms, in order to achieve forest conservation and sustainable development, should be encouraged both domestically and internationally.

    (d) Forest conservation and sustainable development policies should be integrated with economic, trade and other relevant policies.

    (e) Fiscal, trade, industrial, transportation and other policies and practices that may lead to forest degradation should be avoided. Adequate policies, aimed at management, conservation and sustainable development of forests, including, where appropriate, incentives, should be encouraged.

14. Unilateral measures, incompatible with international obligations or agreements, to restrict and/or ban international trade in timber or other forest products should be removed or avoided, in order to attain long-term sustainable forest management.

15. Pollutants, particularly air-borne pollutants, including those responsible for acidic deposition, that are harmful to the health of forest ecosystems at the local, national, regional and global levels should be controlled.

Source: UNCED Information System, Geneva (30 October 1992).

Ongoing Campaigns for Rainforest Protection

In spite of international agreements and resolutions, many organizations worldwide continue to campaign for better protection of rainforests and their resources. On November 18, 1992, the Coalition Against Predatory Logging in the Amazon, which includes nongovernmental organizations (NGOs)—agricultural, environmental, educational, and human rights groups—in Brazil released this manifesto to newspapers and electronic bulletin boards and conferences around the world:

Manifesto to the Population
PREDATORY LOGGING THREATENS AMAZONIA

1. This year marks the five hundredth anniversary of Europeans coming to the Americas. This is the right time for a deep and critical evaluation of the course followed by the societies that devolved from European contact with the peoples and nature of this region. Regarding the relationship of Brazilian society with the environment, the evaluation of these centuries can be defined as a real tragedy. During this period cycles of predatory exploitation of natural resources took place aimed at meeting foreign needs and markets. The consequences of these cycles have been always the same: environmental destruction and social impoverishment.

2. In Brazil the exploitation of Pau Brasil, a red timber and dye, was the first mark of this cruel process. Its consequences are well-known: the disorderly occupation of the territory, the cultural disintegration of indigenous populations, the destruction of forest ecosystems, the extinction of species and the dilapidation of natural resources. This took place for the benefit of a selfish elite and their international partners. (Pau Brasil became commercially extinct in the first century of colonization).

3. Today, five centuries later, we can realize how much this pattern of exploitation, referred to as development, is still present. The Amazon region provides many examples of this pattern of development and is still submitted to a irrational process of devastation and disorderly occupation. Already 415 thousand square kilometres of the Brazilian Amazon have been deforested, about equivalent in size to Germany. The fruits of this devastation are more than questionable: the destruction of tens of indigenous cultures, huge unproductive farms generating very few jobs, illegal and highly polluting gold mining projects, colonists with abysmal living conditions and the advance of prostitution and drug dealing.

Instead of recognizing these mistakes and taking up ecologically viable projects in the areas already deforested (such as agrarian reform, ecological agriculture and forest restoration) the agricultural frontier continues to push into primary forest areas reproducing the same mistakes of the past. As affirmed by the signatories to this Manifesto, colonists prefer to settle in the vast areas already cleared from the forest and do not wish to be pushed into the jungle where living conditions are much harder. It makes no sense to destroy more virgin forest when the result of the deforestation promoted so far are totally absurd in terms of ecological and economic aspects.

4. It is ironic that one of the major economic forces that continues this destructive and disordered penetration of the Amazon forest is very

similar to the activity that extinguished Pau Brasil in the beginning of the colonization of Brazil. Today, the logging industry, and especially Mahogany logging, seriously threatens the future of much of the Amazon forest.

5. Mahogany is the most valuable Brazilian timber sold on the international market. The species is found over a huge area of the southern Amazon covering much of the states of Para, Amazonas, Mato Grosso, Rondonia and Acre. The big sawmill owners who pay for the search and extraction of this "green gold" do not consider how their industry leads to the degradation and eventual complete destruction of primary forest areas. Mahogany is a rare tree and its exploitation requires loggers to move into more and more remote forest areas each year. More than 3,000 kilometres of major roads and tens of thousands of kilometres of secondary roads have been illegally pushed in the southern Amazon in the last ten years to extract Mahogany. After cutting out all marketable Mahogany in a given area the logging companies move on, leaving their access roads for colonists, gold miners and displaced poor who consolidate the destruction of the forest. The exploitation of timber in general and the cutting of Mahogany in particular is the driving force leading to forest destruction in the southern Amazon today.

6. Most of the Mahogany extracted in Brazil is taken from the territories of indigenous peoples. There are numerous reports of Mahogany trees being cut down and removed from indigenous lands despite resistance of indian communities against this usurpation of their territory. At the same time, strong pressure and the allurement of indian leaders—sometimes with the assistance Funai employees— has led some indigenous communities to sign contracts allowing the extraction of Mahogany on their lands. These agreements, however, have not been approved by the relevant government authorities and are legally invalid. Such timber deals have often led to the political fragmentation and cultural degradation of indian groups. These communities learn by example that ecological destruction through over-exploitation of the forest is the only alternative that can allow them access to outside assistance and consumer goods.

In the last ten years the number of indigenous communities that have become victims of the Mahogany boom has more than doubled. The problem is bound to get worse as almost all the remaining stands of Mahogany are located inside indigenous territories. The lack of enforcement from Funai and IBAMA, the federal environment agency, has encouraged the illegal exploitation of Mahogany from indigenous land from logging companies. Over the last two years in the state of Para, the bulk of all Mahogany produced comes from trees illegally extracted from indigenous lands.

7. The ecological impacts of the Mahogany industry are equally serious. Areas of protected forest such as the Biological Reserve of Guapor in Rondonia and the Extractive Reserve Chico Mendes in Acre have been systematically invaded by Mahogany loggers. National Parks throughout the region have also been violated. Due to intense exploitation, the Amazonian Mahogany species, Sweitenia macrophylla King was added in the official list of Brazilian species threatened with extinction in 1992. There is some debate regarding how close the species is to extinction but the fact remains that if left unchecked, the current pattern of exploitation of Mahogany will drive the species to extinction in the Amazon in a few short years.

Wherever the Mahogany exploitation frontier has passed, the species has become practically extinct. In the 'sixties Mahogany was extracted in the Araguaia region of Para. After the commercial extinction of the species in this region, the loggers moved forward along the highway PA 150. During the 'eighties this same predatory pattern moved into the occidental part of Amazonia towards the Xingu River. Today the Mahogany loggers have already crossed the Xingu and keep advancing.

In addition to the role that logging roads play in opening up primary forest to other destructive influences, Mahogany logging itself causes considerable ecological damage. Studies show that for each cut tree around 28 other trees are killed and some 1450 square meters of forest are damaged. The few attempts at growing the species in plantations in the Amazon are still in their early stages and only occupy a small area when compared with the amount of forest damaged by the industry each year. To date none of the plantations have demonstrated the ability to overcome attacks of the moth (Hypssipella grandella) common in commercial Mahogany plantations. Such plantations are often used to justify the continuation of the Mahogany industry in natural forests rather than a real search for concrete alternatives to forest destruction.

8. Despite all the problems mentioned above there are people and companies who still defend the exploitation of Mahogany as a source of economic development for the Brazilian Amazon. Even this argument must be contested. The Mahogany industry is made up of an extensive chain of informal actors and middle-men who are controlled by a small elite group of sawmills and exporters. The industry generates relatively few jobs and the bulk of the profits are made in the importing countries or in the southeast of Brazil. The Mahogany sawmills belong to business groups who moved to Amazonia after exhausting the timber resources in the Atlantic Coast Rainforests and the Araucaria forests in the South of Brazil.

In addition to the impunity with which Mahogany loggers still operate in nature reserves and on indigenous lands, there are strong indications that the industry includes a substantial number of companies engaging in tax fraud through concealing information regarding the source and the correct volume of extracted logs.

9. Given the seriousness of the economic, ecological and social impact of the industry and the clear evidence that Mahogany extraction is perpetuating and intensifying the chaotic model of occupation in Amazonia to the detriment of Brazilian society, the following groups, many of whom deal directly with this problem, have formulated the following demands:

(A) That the Brazilian government through its competent institutions must face up to this problem by prohibiting all cutting and trade of Mahogany in the Amazon region until it has evaluated the extent of damage caused by the industry to date, and defined through a wide debate among all interested parties, legal measures necessary to halt this chaotic process. This measure is necessary to achieve the following objectives:

(a) Mahogany loggers do not build illegal and inadequate roads into primary forest areas, and that existing logging roads are used appropriately or closed.

(b) all exploitation of Mahogany trees in areas designated for ecological preservation is halted.

(c) all exploitation of Mahogany in indigenous areas and extractive reserves is halted and at the same time the government concretely supports the efforts of the forest peoples and communities to find economic and non-predatory alternatives for their survival and development.

(d) the dynamic of predatory logging leading to the extinction of species in areas reached by Mahogany exploitation is stopped.

(B) That the solution of the problems caused by the Mahogany industry must be seen as the first step in the implementation of policies and programmes to end all forms of predatory logging in Amazonia. This policy on Mahogany should serve as a guide to transform all logging activities throughout the region; including restricting the areas where logging is allowed, defining rigidly the technical conditions acceptable for logging operations, halting the violation of protected areas and indigenous territories, and enforcing prohibitions on the cutting of species forbidden by law such as the Brazil-nut tree.

NGOs signing the Manifesto of the "Coalition against Predatory Logging in the Amazon" (November 12):

- Acao Ecologica Vale do Guapore: ECOPORE (Rondonia)
- Associacao de Protecao Ambiental e Recuperacao de Areas Indigenas: APARAI (Rondonia)
- Associacao Profissional dos Engenheiros Florestais do Rio de Janeiro: APEFERJ (Rio de Janeiro)
- AWARU: Organizaao de Apoio ao Povo Nambikwara (Rondonia)
- Casa da Cultura de Maraba (Para)
- Centro Agroambiental do Tocantins: CAT (Para)
- Centro Ecumenico de Documentacao e Informacao: CEDI (Sao Paulo)
- Centro de Educacao,Pesquisa e Assessoria Popular: CEPASP (Para)
- Centro de Trabalho Indigenista: CTI (Sao Paulo)
- Conselho Nacional dos Seringueiros (Acre e Para)
- Centro Mari de Educacao Indigena: CMEI (Sao Paulo)
- Centro de Defesa dos Direitos Humanos (Amazonas)
- Centro Ecumenico de Estudos Biblicos: CEBI (Amazonas)
- Centro dos Trabalhadores da Amazonia: CTI (Acre)
- Centro de Defesa dos Direitos Humanos e Educacao Popular: CDDHEP (Acre)
- Comite Chico Mendes (Acre)
- Comissao pela Criacao do Parque Yanomami: CCPY (Sao Paulo)
- Comissao Pro-Indio do Acre (Acre)
- Comissao Pro-Indio de Sao Paulo: CPI/SP (Sao Paulo)
- Comissao Pastoral da Terra do Acre: CPT/AC (Acre)
- Comissao Pastoral da Terra de Rondonia: CPT/RO (Rondonia)
- Comissao Pastoral da Terra do Amazonas: CPT/AM (Amazonas)
- Comissao Pastoral da Terra do Para e Amapa: CPT/PA (Para)
- Conselho Indigenista Missionario Nacional: CIMI/Nac (Brasilia)
- Conselho Indigenista Missionario do Norte I: CIMI/Norte I (Para)
- Conselho Indigenista Missionario do Norte II: CIMI/Norte II (Amazonas)
- Conselho Indigenista Missionario: Regional Rondonia

- Conselho de Missao entre os Indios da Igreja de Confissao Luterana do Brasil (Rio Grande do Sul)
- Coordenacao das Organizacoes Indigenas da Amazonia Brasileira: COIAB (Amazonas)
- Central Unica dos Trabalhadores do Sudeste do Para (Para)
- Delegacia dos Urbanitarios de Maraba (Para)
- Ecoforca (Sao Paulo)
- Forum das Organizacoes Nao-Governamentais que Atuam em Rondonia (Rondonia)
- Fundacao de Apoio a Vida nos Tropicos—ECOTROPICA (Mato Grosso)
- Fundacao Serra das Andorinhas (Para)
- Fundacao Mata Virgem (Brasilia)
- Fundacao SOS Amazonia (Acre)
- Greenpeace Brasil, Campanha de Florestas (Rio de Janeiro)
- Grupo de Trabalho Missionario Evangelico: GTME (Mato Grosso)
- Instituto de Antropologia e Meio Ambiente: IAMA (Sao Paulo)
- Instituto de Estudos Amazonicos: IEA (Brasilia)
- Instituto de Estudos Socio-Economicos: INESC (Brasilia)
- Instituto de Pesquisa em Defesa da Identidade Amazonica: INDIA (Rondonia)
- Movimento Nacional de Artistas pela Natureza (Brasilia)
- Movimento de Educacao de Base (Para)
- Movimento da Apoio a Resistencia Waimiri-Atroari: MAREWA (Amazonas)
- Nucleo de Direitos Indigenas: NDI (Brasilia)
- Operacao Anchieta: OPAN (Mato Grosso)
- Pastoral Indigenista de Manaus (Amazonas)
- Pastoral Indigenista do Alto Solimoes (Amazonas)
- Protecao Ambiental Cacoalense: PACA (Rondonia)
- Programa de Estudo sobre Terras Indigenas no Brasil: PETI (Rio de Janeiro)
- Sindicato de Pequenos Agricultores e Assalariados Rurais de Rio Branco: SINPASA (Acre)

- Sindicato dos Professores Particulares do Para (Para)
- Sindicato dos Trabalhadores Rurais de Brajo Grande (Para)
- Sindicato dos Trabalhadores Rurais de Breu Branco (Para)
- Sindicato dos Trabalhadores Rurais de Eldorado (Para)
- Sindicato dos Trabalhadores Rurais de Itupiranga (Para)
- Sindicato dos Trabalhadores Rurais de Maraba (Para)
- Sindicato dos Trabalhadores Rurais de Novo Repartimento (Para)
- Sindicato dos Trabalhadores Rurais de Parauapebas (Para)
- Sindicato dos Trabalhadores Rurais de Sao Joao (Para)
- Sindicato dos Trabalhadores Rurais de Tucurui (Para)
- Sindicato dos Trabalhadores em Educacao Publica de Maraba (Para)
- Sociedade Paraense de Defesa dos Direitos Humanos: SDDH (Para)
- Uniao das Nacoes Indigenas do Acre: UNI/AC (Acre)

5

Directory of Organizations

THOUSANDS OF local, state, regional, national, and international organizations participate in projects and programs that are designed to conserve natural resources, including temperate or tropical rainforests, and many new organizations or groups form each year for that purpose. There is space only for a representative sampling of national and international organizations or agencies that sponsor projects or conduct campaigns to protect the global environment or are at the forefront of what has become known as the rainforest movement.

Both general environmental and rainforest groups that emphasize rainforest activities, projects, and programs are included here. Groups that focus on tree planting and restoration of various types of forest areas, but do not necessarily concentrate on rainforest areas—for example, the National Arbor Day Foundation, TreePeople, and Children of the Green Earth—are not listed. These groups as well as others working on diverse environmental issues, including programs for preservation of rainforests, are described in some of the published directories of environmental groups listed in Chapter 6.

American Explorers Club (AEC)
126 Indian Creek Road
Ithaca, NY 14840
(607) 277-0488

This nonprofit scientific and educational organization was founded in 1977 primarily as an information center to advance and support diverse

111

forms of scientific field exploration and research in such areas as biology, geography, and anthropology. Along with its New York headquarters, the AEC maintains offices in Lima, Peru, and Quito, Ecuador. The club disburses contributions for scientific research and exploration and encourages information exchange among scientists and travelers of all nations who explore Central and South America. Another aim of the organization is to awaken greater interest in and appreciation for wilderness conservation and wildlife protection.

PUBLICATIONS: Members receive free the quarterly journal *South American Explorer*. The club also publishes a variety of materials on trip planning and expeditions and a catalog listing a wide variety of maps, books, and crafts.

American Forestry Association (AFA)
P. O. Box 2000
Washington, DC 20013
(202) 667-3300

Founded in 1875, the purpose of the AFA is to maintain and improve the health and value of trees and forests. Through publications and action programs, the organization attempts to interest individuals, industries, and governments in forest conservation. AFA sponsors Global ReLeaf, a major tree-planting effort with projects under way worldwide to improve the earth's environment.

PUBLICATIONS: The bimonthly magazine *American Forests* is free to members and is available to nonmembers for $24 a year. The December 1988 issue focuses on tropical deforestation. The AFA's Forest Policy Center has also published a report on the forest management implications of using the bark of the Pacific yew to produce the anticancer chemical taxol. The yew story shows the value of managing forest ecosystems to protect biodiversity. For a copy of the report, send $5 to Forest Policy Center, American Forestry Association.

Association of Forest Service Employees for Environmental Ethics (AFSEEE)
P.O. Box 11615
Eugene, OR 97440
(503) 484-2692

After 12 years of growing frustration with U.S. Forest Service (USFS) practices, Jeff DeBonis, a timber sale planner, began organizing fellow employees to speak out against agency mismanagement. Thousands of present, retired, and former USFS employees responded and AFSEEE was formed. Its purpose is to "forge a socially responsible value system for the Forest Service based on a land ethic which ensures ecologically and economically sustainable management."

Members of AFSEEE believe that biological diversity, resource sustainability, and the integrity of interrelated ecosystems are more important than managing the U.S. forests primarily for commodity production and political expediency. The membership asserts that timber cutting, grazing, mining, and other resource management activities can be appropriate uses of public lands, but that these activities should not compromise other resource options and values.

PUBLICATIONS: AFSEEE's newspaper, *Inner Voice,* is sent free to members and donors.

Basic Foundation (BF)
P.O. Box 47012
St. Petersburg, FL 33743
(813) 526-9562

Founded in 1970, BF promotes programs around the world that educate the public about the effects of overpopulation, particularly the threat to the world's resources. Through its research efforts, publications, conferences, and nature tours, BF calls for global action to balance population growth with available natural resources. One of BF's most effective educational efforts is organizing nature tours into tropical forests, which in turn prompt support for conservation.

BF also works with Arbofilia in Costa Rica. Arbofilia is a nonprofit organization founded by Costa Ricans. Their goal is "to combine agricultural production with environmental restoration." Arbofilia maintains small nurseries of fruit trees and endangered native timber trees on land owned by subsistence farmers. Members of Arbofilia train the farmers in tree planting and care, and in turn the farmers reforest degraded watersheds and areas not suitable for agriculture.

PUBLICATIONS: BF publishes a brochure about Arbofilia and other educational materials.

Bat Conservation International (BCI)
P.O. Box 162603
Austin, TX 78716-2603
(512) 327-2603

BCI was founded in 1982 to change the negative public perception of bats worldwide and to educate people about the ecological importance of these animals. Bats play a vital role in rainforest rebirth by dispersing seeds and pollinating plants. They also help control vast numbers of crop pests by eating huge amounts of night-flying insects. But because of myths about bats and lack of understanding about their value, many bats have been destroyed and some species are endangered. Membership fees and donations help BCI preserve bats' unique habitats, thus protecting bat populations.

The organization is recognized as the world's leader in bat conservation, and a staff of scientists conducts research on bats, collecting data that contributes to educational programs worldwide.

PUBLICATIONS: The quarterly magazine *Bats* is sent to members. BCI also provides educational materials on bats—publications and audiovisual materials—and a variety of unique gifts described in a catalog that is available on request.

Cascade Holistic Economic Consultants (CHEC)
3758 S.E. Milwaukie
Portland, OR 97202
(503) 234-4349

CHEC is a nonprofit forestry consulting firm that helps people understand and influence the management of public forestlands in the United States. The organization monitors federal forest plans, timber sales, and other forestry programs, analyzes the data, and publishes a variety of educational materials Contributions from Research Patrons help support CHEC.

PUBLICATIONS: CHEC publishes the monthly *Forest Watch* magazine; research papers on forest issues, including "Spotted Owls and Old-Growth"; and Citizens' Guides, such as *The Citizens' Guide to Timber Management* and *The Citizens' Guide to Forestry and Economics*. CHEC economist Randal O'Toole's book *Reforming the Forest Service* (see Chapter 7) may also be ordered from the organization.

Center for Conservation Biology (CCB)
Department of Biological Sciences
Stanford University
Stanford, CA 94305
(415) 723-5924
(415) 723-5920 (FAX)
EcoNet: conbio

Founded in 1984, the CCB is an independent research group associated with Stanford University. The Center focuses on conservation biology, a science that combines the study of population biology and ecology and seeks to apply scientific techniques to the management of populations and ecosystems. One of the Center's goals is to prevent further loss of species and habitats. Its president is Paul Ehrlich of Stanford University.

PUBLICATIONS: The Center publishes a newsletter, *Update*, twice a year.

Conservation International (CI)
1015 18th Street NW, #1002
Washington, DC 20036
(202) 429-5660

Organized in 1987, CI is dedicated to the preservation of tropical eco-systems. CI initiated the debt-for-nature swaps and works in partnership with indigenous peoples to sustain biological diversity and the ecological processes that support life on earth. Current programs are located in Bolivia, Brazil, Colombia, Costa Rica, Guatemala, Madagascar, Mexico, Peru, Suriname, and Venezuela.

PUBLICATIONS: Members receive the newsletter *TROPICUS*. CI also publishes *Orion* magazine, available at many public libraries and for purchase at newsstands.

Cultural Survival (CS)
215 First Street
Cambridge, MA 02142
(617) 621-3818
(617) 621-3814 (FAX)

Established in 1972, CS supports projects that help isolated societies take control of their own destinies and survive contacts with the outside world, which often bring disease, relocation, impoverishment, and even death. CS provides funds and expertise for projects that are implemented by tribal people and ethnic minorities around the world. CS resource management programs enhance the ability of native peoples to manage natural resources. Through its nonprofit marketing division, Cultural Survival Enterprises, the organization establishes markets for products that native communities can harvest in a sustainable manner. A catalog of products is available on request.

PUBLICATIONS: *Cultural Survival Quarterly* magazine is free to members. Cultural Survival also issues news releases about the urgent social and environmental problems faced by smaller societies and publishes research papers and special reports on indigenous populations.

Earth First! (EF!)
Box 7
Canton, NY 13617
(315) 379-9940

"No compromise in defense of Mother Earth" is one of the slogans of Earth First!, which by its own definition (and exclamation point) is a militant but nonviolent ecodefense movement that began about 1980. Those who are part of the movement believe that the wilderness has value for its own sake, not for what it can provide for the benefit of people. It is a concept that stems from Deep Ecology principles emphasizing the interrelatedness of all life. These principles were expressed in earlier times by Henry David Thoreau, John Muir, Aldo Leopold, and others. But EF! goes beyond the theoretical and demands a much tougher stand than traditional environmental groups take to preserve the wilderness, including rainforest areas.

Members use civil disobedience methods or take part in dramatic public protests to prevent destruction of wildlife and wilderness. Some have participated in "monkeywrenching," or vandalizing machinery, cutting fences, and other obstructive tactics to prevent major industries from destroying wilderness areas. Many EF! ecodefense activities are concentrated on forest areas of the Pacific Northwest; they include protesting and sometimes blocking timber companies from logging in national forests. EF! also works with the Rainforest Action Network (see description in this chapter) and other groups working to preserve tropical and temperate rainforests.

PUBLICATIONS: Writings from the now defunct *Earth First! Journal* have been compiled in a book with a foreword by David Foreman, a former leader in the movement.

Earth Island Institute (EII)
300 Broadway, Suite 28
San Francisco, CA 94133
(415) 788-3666
(415) 788-7324 (FAX)
EcoNet: earthisland

In existence since 1982, EII supports international action projects to protect and restore the environment. Projects include educating the public about global warming; coordinating conferences on the interrelationship of environmental protection, economic development, and human rights; informing the public about threats to old-growth forests in the Pacific Northwest; and providing educational materials on rainforest destruction. The organization was founded by David Brower, the Sierra Club's first executive director and the founder of Friends of the Earth and the League of Conservation Voters.

PUBLICATIONS: Members receive the *Earth Island Journal,* a quarterly news magazine.

Earthwatch
P.O. Box 403
680 Mount Auburn Street
Watertown, MA 02272
(617) 926-8200
(617) 926-8532 (FAX)

Earthwatch is an international nonprofit organization established in 1971 to find paying volunteers to help scientists on research expeditions around the world. The mission of Earthwatch is to "improve human understanding of the planet, the diversity of its inhabitants, and the processes that affect the quality of life on Earth." To that end, Earthwatch brings together EarthCorps volunteers and scientists who work on a variety of projects to sustain the world's environment. A number of projects have concentrated on vanishing rainforests and threatened species and their habitats. EarthCorps volunteers share the costs of research expeditions, paying about $1,300 each on average. The Center for Field Research, an affiliate of Earthwatch, reviews proposals from scholars who need volunteer help and screens and selects projects for Earthwatch support.

PUBLICATIONS: *Earthwatch* is the organization's bimonthly magazine that describes the "mission to Earth"—the current expeditions, including project costs and expected field conditions. Earthwatch volunteers receive *The Earth Corps Daily Planet* newsletter, which includes field notes and descriptions of Earthwatch events.

Ecotrust
1200 Northwest Front Avenue
Suite 470
Portland, OR 97209
(503) 227-6225
(503) 222-1517 (FAX)

Ecotrust was established in 1991 to help forge a new strategy for conservation. Affiliated with Conservation International, the organization focuses on the coastal temperate rainforests of North America, helping local people understand these ecosystems and supporting local efforts to develop ecologically and economically sound communities in North American rainforests.

Ecotrust concentrates its efforts in four priority coastal temperate rainforests in North America: the northern Gulf of Alaska, the Gardner Canal and Kitlope River area in British Columbia, the forested islands of Clayoquot Sound, and the Willapa Bay area in southwestern Washington.

PUBLICATIONS: Ecotrust publishes brochures describing its work and has produced a report, *Coastal Temperate Rain Forests: Ecological Characteristics, Status and Distribution Worldwide.*

Environmental Defense Fund (EDF)
257 Park Avenue South
New York, NY 10010
(212) 505-2100
(212) 505-2375 (FAX)

An organization of lawyers, scientists, and economists, the EDF creates innovative, economically viable solutions to environmental problems. Founded in 1967, EDF now has a membership of more than 200,000 and a staff of more than 50. Among the EDF's diverse goals are preservation of tropical rainforests and wildlife. For years, EDF has pressed for changes in the practices of the World Bank and other development banks that lend funds for commercial logging operations in pristine rainforests. In 1991, the World Bank pledged not to fund such projects. Another major EDF project is a four-year traveling exhibition on climate change, with an accompanying book, *Global Warming: Understanding the Forecast,* prepared in conjunction with the American Museum of Natural History.

PUBLICATIONS: Members receive the bimonthly *EDF Letter* and updates on EDF projects. EDF also publishes numerous reports on such topics as acid rain, global warming, recycling, and wildlife conservation.

Food and Agriculture Organization of the United Nations (FAO)
Via Terme di Caaracalla
I-00100 Rome, Italy
(39 6) 57971
Liaison Office for North America
1001 22nd Street, NW
Washington, DC 20437
(202) 653-2402

The FAO is the UN specialized agency responsible for the collection, analysis, and dissemination of information on all aspects of forestry and primary forest products. The Forestry Department, which became a full technical department within the FAO in 1970, also provides technical assistance to developing countries and plays an important role in stimulating forest- and tree-based development.

The Forest Resources Division within the Forestry Department advises the 158 FAO member countries on forestry matters, conducting periodic reviews of international forestry problems and proposing coordinated actions for their resolution. FAO's Committee on Forest Development in the Tropics, which includes representatives from tropical and other countries, studies technical, economic, and social issues relating to the development of tropical forests and reviews international programs in tropical forestry.

PUBLICATIONS: The Forestry Department collates data relevant to forests and forestry worldwide and analyzes and summarizes the data for a variety of publications and issues bulletins, technical and economic reviews, and reports.

Forest History Society (FHS)
701 Vickers Avenue
Durham, NC 27701
(919) 682-9319

Affiliated with Duke University, the FHS was established in 1946 and became a nonprofit educational institution in 1955. The society's purpose is to explore the history of natural resources in the United States and their relationship with people in order to increase public awareness and concern about the nation's forest heritage. In cooperation with other U.S. and Canadian institutions, the society saves historical records related to forests. It also maintains computer databases on forests and provides the information to scholars and others researching forest history. In addition, FHS has produced several films on forestry such as "Timber on the Move: A History of Log Moving Technology" and oral history interviews with former chiefs of the U.S. Forest Service and USFS scientists and planners.

PUBLICATIONS: Members receive the quarterly magazine *Forest & Conservation History.* In conjunction with Duke University Press, FHS also publishes and distributes a number of books on U.S. forestry.

Friends of the Earth (FoE)
218 D Street, SE
Washington, DC
(202) 544-2600
(202) 543-4710 (FAX)
EcoNet: foedc

Founded in 1969, FoE merged in 1990 with the Environmental Policy Institute and the Oceanic Society, forming a single organization, Friends of the Earth. FoE states that it is "an independent, global advocacy organization that works at local, national, and international levels to: protect the planet; preserve biological, cultural, and ethnic diversity; and empower citizens to have a voice in decisions affecting their environment and lives." Among the many projects FoE supports are national and international efforts to stop tropical deforestation and promote forest preservation.

FoE also analyzes public policy from an environmental perspective, advocating research, education, and conservation initiatives. FoE staff routinely monitors activities of the United Nations Environment Program and other UN agencies whose decisions can have an impact on the environment. The organization has called for reform of the Tropical Forestry Action Plan of the UN's Food and Agriculture Organization, urging better protection of the world's rainforests and inclusion of indigenous peoples in planning and administration of new programs.

PUBLICATIONS: Members receive the monthly general newsletter *Friends of the Earth*. FoE also publishes newsletters addressing specific environmental issues such as global climate change.

Global Tomorrow Coalition (GTC)
1325 G Street, NW, Suite 915
Washington, DC 20005
(202) 628-4016
EcoNet: gtc

More than 100 U.S. nonprofit organizations are part of the GTC, which was founded in 1981. GTC is dedicated to educating Americans about the concept of sustainable development and to developing broader public understanding of global trends in population, resources, environment, and development. The coalition sponsors Globescope assemblies in which leaders representing diverse sectors of society in the United States and other nations gather to discuss and confront the most urgent issues of sustainable development. GTC also conducts community forums, congressional briefings, global town meetings, teacher-training workshops, and other special events. In 1992, GTC began a series of "21st Century Dialogues" to bring together representatives of nonprofit organizations and for-profit companies to discuss ways to act on recommendations from the Earth Summit held in June 1992 in Brazil.

PUBLICATIONS: GTC publishes numerous resource materials on sustainable development, such as *The Global Ecology Handbook*; *Sustainable Development: A Guide to Our Common Future*; *Sustainable Development Bibliography*; and *Global Issues Education Packets*. The coalition also produces videotapes on sustainable development.

Greenpeace
Tropical Forests Campaign
1436 U Street, NW
Washington, DC 20009
(202) 462-8817
EcoNet: gp

Greenpeace began in 1971 as a protest against nuclear testing in Alaska and has expanded to become an international organization known for its nonviolent, direct-action campaigns to prevent the emission of pollutants that endanger air, water, and land resources. In 1990, Greenpeace launched its Tropical Forests Campaign, reporting that the campaign would combine "education, consumer organizing, direct action and lobbying [to] pressure government, international financial institutions, corporations and aid agencies to stop contributing to tropical deforestation." The campaign also works with indigenous people of the tropical forests.

PUBLICATIONS: Members receive *Greenpeace,* a quarterly magazine.

International Society for the Preservation of the Tropical Rainforest (ISPTR)
3302 N. Burton Avenue
Rosemead, CA 91770
(818) 572-7273
(818) 572-9521 (FAX)

ISPTR is an American and Peruvian nonprofit organization composed of volunteer naturalists and environmentalists. Since the early 1980s, this dedicated team has been working on the frontline of the Amazon Basin, setting up protected community reserves and wildlife sanctuaries and implementing a series of innovative pilot projects for the benefit of the indigenous people and the fauna and flora of the tropical rainforest. Through one ISPTR project, Preservation of the Amazonian River Dolphin (PARD), the dolphins have been designated as "ambassadors of the rainforest," in an effort to call global attention to the consequences of deforestation and natural habitat destruction.

ISPTR has also collected thousands of pounds of donated clothing, medicine, and educational materials and distributed them among members of Peruvian Indian and mestizo communities, establishing a solid foundation of trust with the local people. In 1990, ISPTR/PARD acquired a 1,000-acre research site on the Yarapa River, a tributary of the Amazon, along which Manuel Cordova harvested many of his medicinal plants (as reported in *Wizard of the Upper Amazon* and *Rio Tigre and Beyond: The Amazon Jungle Medicine of Manuel Cordova,* both by Bruce Lamb (see annotation in Chapter 6). Scientists at a research station on the site conduct multidisciplinary research on tropical rainforest ecosystems. The station also includes an outpost medical clinic for people living in villages alongside the river, and a rehabilitation center for injured, diseased, and orphaned animals that are nursed back to health then released to safe, natural habitats.

ISPTR also established an "Adopt-an-Acre of Rainforest" project to help finance the various programs of the organization, particularly the Cumaceba Communal Reserve (CCR). Funds for the CCR program are used to pay local people to protect the reserve by forming antipoaching teams and river patrols to prevent poaching and injury to animals. Individual students, school classes, and other contributors can participate for $35 per acre and receive a personalized certificate honoring their donation and participation. In addition, ISPTR promotes environmental awareness through a carefully managed ecotourism program and the annual celebration of an International Dedication Day for the Dolphins and Ecology observed each June in Iquitos, Peru, which includes a ceremonial burning of confiscated animal skins to show that animals are more valuable alive than dead.

PUBLICATIONS: Members receive a periodic newsletter, *Amazon Frontline*, describing ISPTR/PARD activities.

International Tropical Timber Organization (ITTO)
Sangyo Boeki Centre Building
2 Yamashita-cho
Naka-ku
Yokohama 231 Japan
(81 45) 671 7045
(81 45 671 7007) (FAX)

The 1983 International Tropical Timber Agreement, under the auspices of the United Nations, established the ITTO in 1985. However, it is an independent intergovernmental organization, operating outside the UN system. Members consist of 22 developing nations that have control of more than 70 percent of the global tropical forests and 24 consuming, or primarily industrialized, countries. Although one of ITTO's functions is to promote trade in timber and other forest products, the organization also promotes research in sustainable timber management and national policies that encourage sustainable development.

PUBLICATIONS: ITTO publishes annual reports and guidelines on sustainable management of tropical forests for timber production.

IUCN-World Conservation Union (IUCN)
Avenue du Mont-Blanc
CH-1196
Gland, Switzerland
022 64 71 81

The International Union for Conservation of Nature and Natural Resources was the former name of this organization, which is an independent coalition of national governments, governmental agencies, and nongovernmental organizations, representing a total of at least 120 countries. The IUCN develops strategies for tackling global environmental problems and conserving natural resources, receiving technical support from a worldwide group of more than 3,000 experts. IUCN objectives include ensuring sustainable development and protection of endangered species. To that end, IUCN monitors conservation efforts, plans conservation strategies, and encourages government agencies and NGOs to initiate conservation programs.

PUBLICATIONS: IUCN publishes program newsletters and a bimonthly *IUCN Bulletin.* IUCN through its Environmental Law Centre in Germany also publishes *Red Data Book,* which describes endangered species.

Lighthawk: The Wings of Conservation
Box 8163
Santa Fe, NM 87504
(505) 982-9656

Known as the "Environmental Air Force," Lighthawk came about in 1978 because of bush pilot Michael Stewartt's idea to let others get an aerial view of environmental damage from industrialization and deforestation. The project began with only one plane and pilot—Stewartt—but now has more than 80 pilots who volunteer their time and their aircraft to take government officials, leaders of environmental groups, journalists, and others on flights over endangered wildlands, including rainforests. Because of Lighthawk flights, the public and policymakers have seen for themselves the kind of devastation that is not visible from the ground. In flights over the Pacific Northwest forests, officials found that the U.S. Forest Service vastly overestimated the amount of ancient forest left in national forests—huge clear-cut patches are visible from the air but do not show on forest service maps. In the same way, Lighthawk flights have spotted—and helped stop—destructive logging practices in British Columbia and illegal logging and mining in Costa Rican rainforests. Lighthawk flights have also helped prompt land conservation projects in Belize.

PUBLICATIONS: Lighthawk publishes a newsletter for members and a brochure describing the organization.

National Audubon Society (NAS)
950 Third Avenue
New York, NY 10022
(212) 832-3200
(212) 593-6254 (FAX)

Biologists, ornithologists, environmental scientists, legislative experts, and more than half a million activist members take part in Audubon projects and programs designed to protect wild species and their habitats in order to guarantee the long-term survival of all species. Founded in 1905, the NAS is one of the oldest conservation organizations in the United States. It has established sanctuaries and nature centers across the nation and has conducted a variety of campaigns to protect wildlife, wetlands, and forests. One high priority campaign is mapping the ancient forests of the Pacific Northwest and lobbying for legislation that will create an ancient forest reserve system. NAS is also involved in a variety of international conservation efforts.

PUBLICATIONS: Members receive the bimonthly *Audubon* magazine. The NAS also publishes *Audubon Activists,* a bimonthly newsletter for members who become part of the Audubon Activist Network. Publications on international issues include materials on rainforests. A list of Audubon resource materials is available on request.

National Wildlife Federation (NWF)
1400 16th Street, NW
Washington, DC 20036-2266
(202) 797-6800
EcoNet: nwfdc

Founded in 1936, the NWF is dedicated to proper management of natural resources and sponsors numerous conservation and environmental education programs, including Wildlife Camps, NatureQuests, and the annual National Wildlife Week. NWF also operates resource centers set up across the United States to investigate and conduct legal efforts, such as supporting legislation and filing lawsuits intended to preserve wildlife.

PUBLICATIONS: Members receive the bimonthly *National Wildlife* or *International Wildlife.* NWF also publishes two magazines for children, *Your Big Backyard* and *Ranger Rick,* and a science and nature activity series called *NatureScope,* which teachers may use as a curriculum supplement.

Natural Resources Defense Council (NRDC)
40 West 20th Street
New York, NY 10011
(212) 727-2700
EcoNet: nrdc

Since 1970 when the NRDC was first initiated by several law school graduates, the organization has worked to protect America's endangered natural resources and to improve the human environment through a combination of scientific research, legal action, and public education. The NRDC frequently provides legal assistance and representation for various environmental causes, such as protecting rainforests and banning the use of pesticides that pose health threats. A major project has been the Rainforest Rescue Campaign to prevent the destruction of tropical and temperate rainforests with special emphasis on preserving Hawaii's endangered rainforest and the ancient forests of the Pacific Northwest.

PUBLICATIONS: Members receive the *NRDC Newsletter* five times a year and the quarterly magazine, *The Amicus Journal,* a name that comes from the Latin phrase *amicus curiae* (friend of the court), a role that NRDC often plays on behalf of the environment. NRDC also published *Rainforest Book* (see Chapter 6 annotation) and materials on dangerous pesticides. A complete list of NRDC's books and reports is available on request.

New Forests Project (NFP)
731 8th Street, SE
Washington, DC 20003
(202) 547-3800
(202) 546-4784 (FAX)

The nonprofit International Center, which promotes democratic movements in developing countries, established the NFP in 1982 to help provide means for restoring degraded soil caused by deforestation and to improve self-reliance. With the emphasis on self-help, NFP has set up demonstration sites where field workers show local people how to plant, wisely manage, harvest, and use fast-growing trees. Through the training program, people learn to rotate their harvest to assure sustained yield and to prevent soil erosion by way of tree plantings.

According to NFP, the organization focuses on "fast-growing, nitrogen-fixing tree species that have the potential for multiple uses. With proper management, these trees can sustainably produce fuelwood, livestock feed, and organic fertilizer. In addition to being excellent choices for soil rehabilitation and erosion control projects, these trees have a proven ability to increase agricultural yields."

Since its inception, NFP has helped local farmers begin tree-planting projects in several thousand villages in over 100 developing countries. NFP receives assistance from the World Seed Program and support from members and donors.

PUBLICATIONS: The quarterly newsletter *New Forests News* is sent free to members and supporters.

Oregon Natural Resources Council (ONRC)
Yeon Building, Suite 1050
522 Southwest Fifth Avenue
Portland, OR 97204
(503) 223-9001

Founded in 1973, the ONRC is a coalition of more than 40 conservation, recreation, commercial, educational, and sports groups concerned about the wise management of Oregon's natural resources. Although it is a state organization, its efforts have broad impact. Leaders of the ONRC lobby U.S. Congressional members to enact legislation that will protect the Pacific Northwest ancient forests along the "emerald string" and to ban exports of timber from old-growth forests. Other ONRC efforts include campaigns to protect the northern spotted owl and other endangered species and to disseminate information about destructive timber sales and unsustainable forest management practices.

PUBLICATIONS: The quarterly *Wild Oregon* is sent to members. ONRC publishes bulletins alerting members to actions they can take and also educational materials on Oregon's natural resources.

The Population Institute (PI)
107 Second Street, NE
Washington, DC 20002
(202) 544-3300

"No matter what your cause—it's a lost cause—if we don't come to grips with overpopulation." That is a basic theme of PI, founded in 1969 to promote stabilization of the world's population, which now stands at 5.4 billion. In contrast, the institute points out, it took all of recorded history for world population to reach one billion by 1830. At the current rate of growth, the world population will grow by more than 95 million each year, and the total population will double in fewer than 40 years, placing great demands on limited natural resources.

One of PI's main goals is to educate the public about the links between rapid population growth and short-term government policies designed to provide for increasing numbers of people but instead contributing to environmental catastrophes, such as deforestation, soil erosion, loss of plant and animal species, and global warming. PI also stresses the U.S. role in overpopulation issues, pointing out that "the United States with its ever growing population and over consumption habits poses a much more immediate and larger threat to the environment than that of developing nations."

PI's communication and education efforts have concentrated on family planning assistance, particularly in developing countries where millions of women want to plan the size of their families but lack the means and access to reliable birth control. A major campaign is the Population/Environment effort to persuade nations to increase support for the UN Population Fund and to encourage the work of voluntary family planning organizations.

PUBLICATIONS: PI publishes a bimonthly *Popline,* which contains news and features designed for use in major newspapers worldwide. The organization also publishes many other public education materials, including a monograph series on a variety of population/environment issues, brochures, and reports. A list of PI materials and an order form are available on request.

Rainforest Action Network (RAN)
301 Broadway, Suite A
San Francisco, CA 94133
(415) 398-4404
EcoNet: ran

In 1985, RAN was set up to work internationally with other environmental organizations on major campaigns to protect rainforests and the human rights of those living in and around those forests. The network convened the first international rainforest conference, which brought together activists from 35 organizations to develop action plans that helped catalyze the world movement to protect rainforests.

One of RAN's first action campaigns was a boycott of the Burger King restaurant chain to protest the company's practice of importing inexpensive beef from tropical countries where cattle ranching was destroying rainforests. Rainforest Action Groups (RAGs) formed to conduct the campaign and have continued since then to work at the grassroots level to draw public attention to human activities that result in massive deforestation. RAN and RAGs have focused on the tropical timber trade and have called for a ban on the import and consumption of tropical timber products in the United States and other industrialized nations.

PUBLICATIONS: The quarterly news magazine *World Rainforest Report* is free to members as is the monthly *Action Alert*. Information about threats to rainforests is also published electronically, communicated via computer networks around the world.

Rainforest Alliance (RA)
270 Lafayette Street, Suite 512
New York, NY 10012
(212) 941-1900
(212) 941-4986 (FAX)

The primary mission of RA is "to develop and promote economically viable and socially desirable alternatives to tropical deforestation." To that end, RA helps with the development of forest products in cooperation with indigenous groups and other local organizations. Projects provide long-term stable income for people living in or near tropical forests. One major activity is the Smart Wood Certification Program, which awards certificates to well-managed sources of tropical woods. Sources receiving the certification have met strict standards that ensure forests are managed in nondestructive ways and that local peoples receive long-term benefits. Other activities include promoting sustainable cultivation and harvesting of medicinal plants; conducting fundraisers that benefit debt-for-nature swaps; supporting field research in tropical forests; providing grants to community programs struggling to earn a livelihood without harming tropical forests or their wildlife.

PUBLICATIONS: *The Canopy,* a quarterly newsletter, is free to members. The organization also offers curriculum packages for nominal fees and supports a Tropical Conservation Newsbureau based in Costa Rica. The staff of the Newsbureau writes and distributes feature articles describing successful conservation efforts, to 500 media outlets in the United States as well as Spanish-language versions to 300 media outlets, government institutions, and conservation organizations in Latin America.

Rainforest Defense Fund (RDF)

P.O. Box 2104
Cambridge, MA 02238
(617) 628-9788

Established as a nonprofit organization in 1991, the RDF provides legal defense for rainforest people whose survival is threatened by deforestation. One of RDF's missions is to educate the public about rainforests. Another goal is to provide for industries environmentally safe alternatives to manufacturing products and methods that cause ecological damage.

RDF is currently developing a rainforest database, including locations and information on indigenous groups, and will make this information available to environmental organizations. Among other efforts under way is development of a guidebook for activist groups and a program to place volunteers with indigenous groups in endangered rainforest areas.

PUBLICATIONS: Members receive a monthly newsletter describing actions they can take.

Save America's Forests (SAF)

4 Library Court
Washington, DC 20003
(202) 544-9219

Founded in 1990 as a nonprofit organization, SAF calls itself a "nationwide coalition of grassroots environmental groups, public interest organizations, responsible businesses and individuals working to pass strong, comprehensive nationwide laws to protect our forest ecosystems." One purpose of the organization is to bring grassroots activists to Washington, D.C., to testify before the U.S. Congress and to help educate members of Congress about the principles of biodiversity and the need to change the way national forests are managed. SAF hopes to counteract the intensive efforts of the timber industry, which has successfully campaigned for laws that favor the industry over the environment.

PUBLICATIONS: SAF publishes periodic reports for members.

Sierra Club (SC)

730 Polk Street
San Francisco, CA 94109
(415) 776-2211

Naturalist John Muir founded the SC in 1892, and since then the environmental organization has established chapters throughout the United States with at least 500,000 members promoting the responsible use of the earth's ecosystems and resources. Through its various litigation and public education programs, the club has helped establish national parks and wilderness preserves. It also provides information and encourages grassroots support for a variety of programs to protect the environment, ranging from efforts to stop acid rain to reducing deforestation.

PUBLICATIONS: The quarterly magazine *Sierra* is free to members. SC also publishes numerous books on environmental topics, news reports and releases, and newsletters for members of chapters and local groups.

Smithsonian Institution
Washington, DC 20560
(202) 357-2627
(202) 786-2377 (FAX)

Along with preserving artifacts and works of art that reflect America's cultural heritage, the Smithsonian, which was founded in 1846, sponsors scientific research and programs to protect the environment. The Smithsonian's National Museum of Natural History has initiated a Biodiversity Program, which includes research programs in Latin America (BIOLAT) and a Biological Dynamics of Forest Fragments Project (BDFF) in Amazonia. BIOLAT is designed to conduct basic research on biodiversity and support scientific projects in the field. The basic purpose of BDFF is "to identify a minimum size of tropical foret habitat that would maintain most of the biotic diversity represented in an intact ecosystem." BDFF researchers study forest plants and animals before and after they have been fragmented by cattle ranches and compare the data gathered to determine what size forest area is needed to maintain the original biodiversity.

PUBLICATIONS: Members receive the monthly *Smithsonian Magazine*. The Smithsonian Institution Press publishes a wide variety of books, technical papers, exhibition catalogs, and educational materials.

The Student Conservation Association (SCA)
P.O. Box 550
Charlestown, NH 03603
(603) 826-4301
(603) 826-7755 (FAX)

Through the SCA, students and others looking for work experience and educational opportunities can volunteer their services for conservation projects in national parks and other natural resource areas that belong to the public. SCA works with such federal agencies as the National Park Service, the U.S. Forest Service, the Bureau of Land Management, and state park and wildlife agencies. Each year, the SCA places about 1,300 volunteers as resource assistants in conservation programs across the United States. Volunteers usually receive a travel grant to pay for transportation to their program area, free housing at the conservation site, and an allowance for food and basic living expenses.

PUBLICATIONS: Members receive a monthly *JOB-SCAN*, which lists paid employment opportunities in resource management.

Survival International (SI)
310 Edgeware Road
London, England, W2, 1DY
(071) 723-5535

This British-based group, with members in 63 countries, campaigns for the rights of threatened tribal peoples, helping to ensure their rights to self-determination and control of natural resources. SI pressures oppressive governments to protect tribal rights and opposes environmentally unsound development projects.

PUBLICATIONS: Members receive *Survival International News*. SI also publishes *Urgent Action Bulletins* and other educational materials.

Threshold, Inc.
Drawer CU
Bisbee, AZ 85603
(602) 432-7353

Established in 1972, Threshold is a nonprofit organization that, since its inception, has been one of the primary forces in creating global awareness of tropical rainforest destruction and has funded numerous projects in environmental protection, research, technical assistance, and public education. Working at regional, national, and international levels, Threshold projects include support for rubber tappers and indigenous tribal peoples of the Amazon Basin and a variety of tropical forest action groups in Asia and the Pacific region; educational efforts such as a video "Seeds of Hope" and a traveling media exhibit on Asian tropical forests that have been preserved through grassroots organizations; and international conferences that help develop coalitions and environmentally sound development planning in tropical countries.

Threshold recently established The Environmental Crisis Fund to "help heal the major ecological threats now affecting planet Earth." Committed to protecting critically endangered ecosystems, the fund is designed to rush needed financial support directly to local environmental action and study groups and to develop coalitions of groups working to solve environmental problems. Currently Threshold's Environmental Crisis Fund is developing or exploring the possibility of projects that will assist in protection of the Yanomami Indian tropical forest homelands in Venezuela, preservation of tropical ecosystems in Madagascar and the endangered tropical forests in Hawaii.

PUBLICATIONS: Threshold publishes a brochure about its activities and a variety of educational materials on global environmental issues.

Trees for the Future (TFF)
11306 Estona Drive
P.O. Box 1786
Silver Spring, MD 20915-1786
(301) 929-0238

Founded in 1979, the nonprofit TFF has been establishing self-help projects worldwide designed to protect the world's remaining forests and bring life back to lands that have been destroyed by erosion, flooding, desertification, and infertility. TFF concentrates on planting fast-growing trees that can be harvested for fuelwood and building materials and help restore impoverished lands.

TFF technicians are providing the means to help villagers in Africa, Asia, and Latin America plant more than a million trees every month. TFF also has established seed production farms in 11 developing countries.

PUBLICATIONS: TFF publishes a brochure explaining the program and regular updates on the tree-planting efforts.

United Nations Environment Programme (UNEP)
P.O. Box 30552
Nairobi, Kenya
(254 2) 333930
New York Liaison Office
DC2-0803 United Nations
New York, NY 10017
(212) 963-8093

Through its specialized agencies, the United Nations conducts a number of programs to help protect the global environment and manage natural resources. UNEP was established in 1972 as a result of the UN Conference on the Human Environment in Stockholm, Sweden. Although UNEP is not directly involved in specific environmental projects, it oversees the work of other UN agencies, coordinating and stimulating action in such areas as climate change, desertification control, water quality, and deforestation.

PUBLICATIONS: UNEP publishes an annual report, *The State of the World Environment,* and numerous environmental documents, newsletters, guidelines, reports, and other materials. UNEP publications are listed in the organization's annual report.

Western Ancient Forest Campaign (WAFC)
1400 16th Street NW, Suite 294
Washington, DC 20036
(202) 939-3324

Created and launched in 1991, the WAFC is a network of advocates for conservation of America's endangered ancient forests. A nonmembership, nonprofit organization, its purpose is to expand and activate the citizen conservation community to educate the American public and policymakers on the value of protecting U.S. ancient forest ecosystems. WAFC's Washington staff and regional coordinators in the Pacific Northwest educate and organize grassroots activists who campaign for federal legislation that will protect the remaining ancient forests in the United States.

The three major program areas of the campaign are advocacy, grassroots involvement, and public education. Distributing information is WAFC's "single most important role," and the network uses fax machines, direct mail, online computer conferences, and legislative "hot lines" to get out the word on the latest actions in the courts, federal agencies, and Congress regarding ancient forests.

PUBLICATIONS: WAFC publishes *Report from Washington,* brochures, news releases, and similar materials to publicize the need to protect America's ancient forests.

The Wilderness Society (WS)
900 17th Street, NW
Washington, DC 20006
(202) 833-2300

Devoted to preserving wilderness and wildlife, the WS has established programs to protect America's prime forests, parks, rivers, and shorelands. Since its founding in 1935, WS has worked for federal legislation that sets aside wilderness areas for preservation. Among its efforts to conserve wilderness areas throughout the United States, WS has been instrumental in measures to help protect the Tongass National Forest in Alaska.

PUBLICATIONS: Members receive *The Wildlifer,* a bimonthly newsletter. The society also publishes the quarterly *Journal of Wildlife Management* and *The Wildlife Society Bulletin.*

World Forestry Center (WFC)
4033 S.W. Canyon Road
Portland, OR 97221
(503) 228-1367

"Inform, involve, inspire" is the motto of the WFC, founded in 1971 to educate the public about the importance of well-managed forests. The Center has created special exhibits on forestry and forests of the world for its indoor and outdoor exhibits and demonstration tree farm/forest near Portland. The WFC prepared an exhibit "Old Growth Forests: Treasure in Transition" for showing at the Smithsonian Institution in Washington, D.C., and produces numerous educational materials on forestry and conservation of natural resources. A branch of WFC, the World Forestry Institute, gathers and circulates data and research on forestry worldwide.

PUBLICATIONS: Members receive the bimonthly newsletter *Branching Out*. The organization also publishes a quarterly magazine *Forest Perspectives: New Directions in Natural Resource Management* and various teacher's guides. WFC also offers a Rainforest Kit and an Old-Growth Forest Kit, as well as a kit on urban forestry for a two-week rental fee of $25 plus a refundable deposit of $25. The kits include posters, videos, and other educational materials.

World Resources Institute (WRI)
1709 New York Avenue, NW
Washington, DC 20006
(202) 638-6300

WRI was founded in 1982 and later merged with the Center for International Development and Environment. WRI is a policy research center that assists governments, international organizations, and the private sector with basic policy decisions in regard to balancing human needs and economic growth with preservation of natural resources. The institute focuses on such issues as biological diversity, tropical forests, and incentives for sustainable development. In nonindustrialized nations, WRI provides technical assistance for government agencies and NGOs working to manage resources sustainably.

PUBLICATIONS: WRI publishes policy studies, research reports, and dozens of other materials on the world's resources. An annual reference work, *World Resources*, provides information on global resources by nation.

World Wildlife Fund (WWF)
1250 24th Street, NW
Washington, DC
(202) 293-4800

Founded in 1961, the WWF is the largest conservation organization working worldwide to protect endangered wildlife and wildlands. WWF is dedicated to "reversing the degradation of our planet's natural environment and to building a future in which human needs are met in harmony with the international WWF network." A number of WWF conservation projects are under way in the tropical forests of Latin America, Asia, and Africa. For example, in the state of Para in eastern Amazonia, WWF is helping several communities find alternatives to the slash-and-burn pattern of agriculture. Local people are beginning to harvest valuable products and gain a fair return for their goods, which helps motivate preservation of the rainforest. WWF has helped preserve nearly 200 national parks and reserves.

PUBLICATIONS: Members receive the bimonthly newsletter *Focus*. The organization also publishes *Wildlife Alerts*, describing current efforts to save endangered species.

Worldwatch Institute (WI)
1776 Massachusetts Avenue, NW
Washington, DC 20036
(202) 452-1999
(202) 296-7365 (FAX)

The well-known and respected WI was founded in 1974 to "inform policymakers and the public about the complex links between the world economy and its environmental support systems." WI analyzes data from hundreds of research sources, including scientists and international organizations, and disseminates information through major media around the world.

PUBLICATIONS: The institute publishes an annual *State of the World* report; a variety of *Worldwatch Papers* on major environmental issues; the *Worldwatch* magazine; *The World Watch Reader*, an anthology of *Worldwatch* magazine articles; and an Environmental Alert Series that includes two or three publications per year on basic environmental problems. A catalog of publications is available on request.

Zero Population Growth (ZPG)
1400 16th Street, NW, Suite 320
Washington, DC 20036
(202) 332-2200

Paul Ehrlich, coauthor of *The Population Explosion* and author of an earlier work, *The Population Bomb*, founded ZPG in 1968. The philosophy behind ZPG and its purpose are outlined in these books. Basically, this nonprofit organization works worldwide to bring about a balance among population, resources, and the environment. To that end, ZPG produces numerous educational materials, focusing on diverse but interrelated issues such as voluntary family planning, sustainable development, and U.S. foreign aid legislation.

PUBLICATIONS: ZPG sends the *ZPG Reporter* to members free and also publishes a variety of newsletters, fact sheets, and brochures.

6

Selected Print Resources

SINCE THE 1960s, an increasing number of books have been published on rainforest topics, with a proliferation of titles appearing during the 1980s and early 1990s. Because there are hundreds of titles on rainforests or related topics, selections for annotation were limited primarily to those published since the mid-1980s.

Many of the books listed in this chapter were written for a general readership; some are for readers who are looking for technical information and data on the world's rainforests, and others are for conservationists and environmental activists. Whatever the background of readers, those interested in temperate or tropical rainforests should be able to find references suited to their particular purposes. Atlases, anthologies, directories, and guides that contain information on rainforests make up the first part of the listing. Following this section is an alphabetical listing of books on a variety of subjects pertinent to the study of rainforests: Biodiversity and Species Extinction; General Environmental Subjects with sections on rainforests or information related to rainforests; a few titles on Global Warming that have sections explaining how climate change is related to deforestation; Indigenous People; a listing of books on Sustainable Society and the role of rainforest preservation in sustainability; Temperate Rainforests; and, finally, Tropical Rainforests.

Atlases, Anthologies, Directories, and Guides

Ashworth, William. **The Encyclopedia of Environmental Studies.** New York: Facts on File, 1991. 470p. ISBN 0-8160-1531-7.

This comprehensive reference work is an excellent resource for anyone interested in environmental issues, including rainforest protection. The encyclopedia contains more than 3,000 entries defining environmental terms from many diverse disciplines, such as biology, botany, chemistry, economics, geography, and geology. Entries also describe individuals who have had an impact on the environmental movement from Edward Abbey to James Watt. Major U.S. environmental laws, regulations, and regulatory agencies, and significant events that led to environmental protection are explained. Diagrams, tables, and cross-references throughout enhance understanding of concepts and terms. A thorough bibliography is included.

Boo, Elizabeth. **Ecotourism: The Potentials and Pitfalls** (2 vols). Baltimore, MD: World Wildlife Fund (WWF), 1990. Vol. 1, 85p.; Vol. 2, 123p. ISBN 0-942635-16-7 (set).

Focusing on Latin America and the Caribbean, these books evaluate the economic and environmental impacts of ecotourism. Country case studies are included in the second volume.

Brainard, John C., ed. **The Directory of National Environmental Organizations.** St. Paul, MN: U.S. Environmental Directories, 1992. 195p. [no ISBN]

This directory lists more than 600 environmental and conservation organizations, with addresses, phone numbers, and descriptions. Each entry includes the number of members and the date the organization was established. There is a subject index as well as a geographic index that shows locations of organization headquarters and federal environmental agencies.

Brown, Lester R., and the Worldwatch Institute. **State of the World: A Worldwatch Institute Report on Progress toward a Sustainable Society.** New York: W.W. Norton, (annual since 1984). 250p. ISBN 0-393-03439-9.

In its ninth edition as of 1992 and published in 26 languages, the **State of the World** is the most widely used analysis of public policy in the world. This desktop guide presents the results of an "annual physical" of the world, including chapters relating to rainforest issues, and includes more than 50 figures and tables.

Burton, John, ed. **The Atlas of Endangered Species.** New York: Macmillan, 1991. 256p. ISBN 0-02-897081-0.

Written by a team of experts, this atlas describes and illustrates numerous animal and plant species around the world in danger of extinction. With color illustrations and maps, the atlas is organized by regions and includes various focus sections, including one on tropical timber that describes endangered timber resources in Africa, the Philippines, and Latin America. The atlas also has a section on "Conservation in Action," lists of endangered wildlife, a directory of conservation organizations, and an extensive bibliography.

Buzzworm Editors. **1993 Earth Journal.** Boulder, CO: *Buzzworm* Magazine, 1993. 448p. ISBN 0-9603-722-9-6.

This is the second annual edition of an environmental almanac and resource directory prepared by the editors of *Buzzworm* magazine. The journal includes an "Earth Diary" covering environmental events of the previous year; reports on environmental issues by world experts; a section on "Ecoculture" describing Green Businesses, EcoTravel, and The Environmental Home; and "Ecovoice," a section of environmental commentary and essays.

Castner, James L. Foreword by Peter H. Raven. **Rainforests: A Guide to Research and Tourist Facilities at Selected Tropical Forest Sites in Central and South America.** Gainesville, FL: Feline Press, 1990. 350p. ISBN 0-9625150-2-7.

According to the author, "This book was written for anyone who has the desire to visit a rainforest for whatever reason." Castner describes a variety of tropical forest sites, and the book serves as a guide for those who want an intimate look at a rainforest community. It includes a chapter on how to find funding for research in tropical biology and an annotated bibliography of more than 200 references on tropical rainforests.

Collins, Mark, ed. **The Last Rain Forests: A World Conservation Atlas.** Emmaus, PA: Rodale Press, 1990. 200p. ISBN 0-19-520836-6.

Prepared in collaboration with the International Union for Conservation of Nature and the World Monitoring Centre, this atlas of more than 50 rainforests worldwide includes photographs of flora and fauna and shows how the forests are being destroyed. The text describes the economic problems of rainforest people. Also included is an explanation of measures that rainforest nations and international conservation groups are taking to conserve the forests.

Collins, Mark N., Jeffery Sayer, and Timothy C. Whitmore. **The Conservation Atlas of Tropical Forests: Asia and the Pacific.** New York: Macmillan, 1991. 256p. ISBN 0-131-79227-1.

One of the few atlases on Asian and Pacific tropical forests, this volume contains a treatment of each country in the regions from India to the Western Pacific Islands. Along with large, clearly detailed maps of botanic gardens and conservation areas, numerous statistics are included in this book.

Emmons, Louise H., and Francois Feer. **Neotropical Rainforest Mammals: A Field Guide.** Chicago: University of Chicago Press, 1990. 550p. ISBN 0-226-20716-1 (cloth); 0-226-20718-8 (paper).

Illustrated with beautiful photographs by Francois Feer, this field guide is a first of its kind. It is a directory of mammals one might encounter in Central and South America at elevations below 1,000 meters and contains information about approximately 500 species, including measurements and markings, geographic variation, natural history, and geographic range.

Gennino, Angela, ed. **Amazonia: Voices from the Rainforest.** San Francisco: Rainforest Action Network, 1990. 92p. ISBN 0-9628033-0-8.

A highly recommended reference, this paperback guide lists and provides profiles of 250 international organizations working on Amazonia issues. Groups are active in Latin America, Europe, North America, and the Asia/Pacific area. The reference also includes a bibliography of recommended books and films on Amazonia. It is a good resource for anyone interested in the politics of deforestation and efforts worldwide to stop the destruction of rainforests.

Goldsmith, Edward, and Nicholas Hildyard, eds. **The Earth Report: The Essential Guide to Global Ecological Issues.** Los Angeles: Price Stern Sloan, 1988. 240p. ISBN 0-89596-673-0.

A handbook on environmental issues, this volume includes essays and an A-to-Z encyclopedia of ecological concepts, terms, and facts. Entries cover topics ranging from deforestation in the Amazon to the Gaia hypothesis to the World Bank. The guide is illustrated with clear charts, diagrams, and photographs.

Gorder, Cheryl. **Green Earth Resource Guide.** Tempe, AZ: Blue Bird Publishing, 1991. 254p. ISBN 0-933025-23-8.

This comprehensive guide is primarily a listing of businesses and products that do not harm the environment, but it also includes chapters relating to rainforest preservation, such as those on ecotourism and recycled paper products. Among the environmental organizations listed are those working to protect rainforests.

Holing, Dwight. **earthTrips: A Guide to Nature Travel on a Fragile Planet.** Venice, CA: Living Planet Press, 1991. 209p. ISBN 1-879326-05-1.

A Conservation International book, this guide is about ecotourism, an increasingly popular way to travel, which is, in the author's words, "ecologically sensitive travel that combines the pleasures of discovering and understanding spectacular flora and fauna with an opportunity to contribute to their protection." By attracting tourists to national parks and other reserves, many of which are rainforests, local people are able to earn income and at the same time protect natural areas. Another purpose of ecotourism is to promote environmental awareness and conservation worldwide. This book explains how to "travel with a cause" in seven regions of the world, describes volunteer research and nature-study vacations, and lists "environmentally responsible travel organizations."

Lanier-Graham, Susan D. **The Nature Directory: A Guide to Environmental Organizations.** New York: Walker and Company, 1991. 190p. ISBN 0-8027-7348-6.

From the African Wildlife Foundation to Zero Population Growth, prominent environmental organizations are described in this directory. Each entry includes a brief history and goals of the organization, descriptions of past achievements, ongoing projects, future plans, and membership information. The guide also includes suggestions for further reading on environmental problems.

Mason, Robert J., and Mark T. Mattson. **Atlas of United States Environmental Issues.** New York: Macmillan, 1990. 252p. ISBN 0-02- 897261-9.

Described as a research and teaching tool, this atlas can be a primary source for students and others investigating environmental issues in the United States. It includes four-color maps, charts, graphs, and diagrams. The text covers such topics as agricultural lands; forests and forestry; parks, recreation, and wildlife; the environment and politics; and major environmental legislation.

Middleton, Nick. **Atlas of Environmental Issues.** New York: Facts on File, 1989. 63p. ISBN 0-8160-2023-X.

With 28 sections on environmental topics, this reference book includes brief information on deforestation, trading in endangered species, threatened species, and wildlife tourism. Color photographs and maps supplement the text.

National Wildlife Federation. **1992 Conservation Directory.** Washington, DC: National Wildlife Federation, 1992. 398p. ISBN 0-945051-51-4.

Revised each year, this directory lists governmental and nongovernmental organizations engaged in conservation efforts at the state, national, or international level. Entries include titles and names of personnel. There are also brief descriptions of agencies and organizations and an extensive index by publication and subject.

Seager, Joni. **The State of the Earth Atlas.** New York, London, and Tokyo: Simon & Schuster, 1990. 127p. ISBN 0-671-70524-5.

Combining full-color maps and charts with text, this atlas shows clearly how human activities have made an impact on the global environment. Various sections cover such topics as Tropical Forests, Fossil Fuel Pollution, The Timber Trade, and an international table of statistics by country.

Seredich, John, ed. **Your Resource Guide to Environmental Organizations.** Irvine, CA: Smiling Dolphins Press, 1991. 514p. ISBN 1-879072-00-9.

This well-organized and easy-to-read guide describes 150 environmental organizations, including many involved in rainforest preservation work. Closeups of 14 environmental leaders are also part of the work, as is a glossary.

Stein, Edith C. **The Environmental Sourcebook.** New York: Lyons & Burford, 1992. 264p. ISBN 1-55821-164-0.

Written by a physician who specializes in environmental medicine and is president of the Environmental Data Research Institute, this sourcebook is designed to help citizens understand major environmental issues, such as rainforest destruction, endangered species, and atmospheric pollution. The book also explains which foundations fund environmental causes and where readers can find further information on a particular ecological issue. Much of the information in this volume is also available in an electronic database, as described in Chapter 7.

Trzyna, Thaddeus C., ed. **World Directory of Environmental Organizations,** 3d ed. Claremont, CA: California Institute of Public Affairs, 1989. 167p. ISBN 0-912102-87-X.

This is a handbook of national and international organizations and programs of governments and nongovernmental organizations concerned with the protection of the earth's resources. It was prepared in cooperation with the Sierra Club and the World Conservation Union and is a completely rewritten version of earlier (1973 and 1976) editions.

Wild, Russell, ed. **The Earth Care Annual.** Emmaus, PA: Rodale Press, 1993. 235p. ISBN 0-87596-136-3.

Published since 1990, this annual is a collection of articles about environmental topics and people who are active in projects to protect the earth. Editions to date contain articles about activists, including indigenous people, in rainforest preservation.

The World Resources Institute. **World Resources 1992-93.** New York and Oxford: Oxford University Press, 1992. 383p. ISBN 0-19-506230-2 (cloth); 0-19-506231-0 (paper).

A biennial report prepared in collaboration with the UN Environment Programme and the UN Development Programme, this volume follows the format and builds on earlier editions of **World Resources.** It features a special section on sustainable development (Part I) and covers conditions and trends in such areas as population and human development, forests and rangelands, wildlife and habitat, and atmosphere and climate. A new analysis of nongovernmental organizations is part of this edition as are reports on the first global survey of land degradation. An updated report on tropical deforestation is also included. Part IV consists of more than 130 pages of data tables.

Biodiversity and Species Extinction

Bates, Marston. **The Forest and the Sea.** New York: Lyons & Burford, 1988. 288p. ISBN 1-55821-009-1 (paper).

This is a reprint of a classic work on the general principles of the organization of the biological community. It examines the patterns of relationships among individuals, populations, and species. A naturalist, Bates eloquently describes the different biomes: the seas, the rivers, the rainforests, the woodlands, and the deserts. He places humans in this system as a natural part of it rather than a special phenomenon of it. Endnotes are included.

Ehrlich, Paul R., and Anne H. Ehrlich. **Extinction—The Causes and Consequences of the Disappearance of Species.** New York: Random House, 1981. 294p. ISBN 0-394-51312-6.

The Ehrlichs show with strong, reasoned arguments and documentation how the extinction of just a single species could lead to a disaster. They explain how humans have endangered certain species, how they benefit from endangered species, and what can be done to protect them.

Hoage, R. J. **Animal Extinctions: What Everyone Should Know.** Washington, DC: Smithsonian Institution Press, 1985. 191p. ISBN 0-87474-521-7.

This is a collection of 12 papers that were presented at the first National Zoological Park Symposium for the Public in 1982. Of interest are the papers that explore the global implications of species extinction, specifically a study on Barro Colorado Island and a paper on scientific strategies for preserving species focused on the Brazilian rainforest, although all the papers can be applied to Amazonia.

Huxley, Anthony. **Green Inheritance—The World Wildlife Fund Book of Plants.** New York: Anchor/Doubleday, 1985. 193p. ISBN 0-385-1903-2.

Although this book is not specifically about rainforests, it explains the importance of plants in our world and why we must work to save them. Huxley explains that plants are the basis of the food cycle and that if they are the first to disappear they signal the next step: the disappearance of animals, including humans. In many chapters, Huxley points out the importance of rainforest plants and their products. Individual chapters include discussions of plants as sustenance, plants as medicine, plants as crops, and plants as objects of beauty. There is also a brief explanation of how the loss of vegetation contributes to the greenhouse effect.

Kaufman, Les, and Kenneth Mallory, eds. **The Last Extinction.** Cambridge, MA and London: The MIT Press, 1986. 208p. ISBN 0-262-11115-2.

Based on a public lecture series called "Extinction: Saving the Sinking Ark" held in Boston, the contributors to this book outline what ecology and species extinction are about. The collection of essays, written by various experts, is designed "to awaken the general public to the issues underlying the notion that we must prevent a 'last extinction' " and even though "mass extinction is in progress . . . it can be postponed indefinitely." The first chapter is an overview of the complex problem of species extinction, and the second reviews evidence of past extinctions. Ongoing destruction of tropical forests is the subject of chapter three, while the fourth essay describes "Vanishing Species in Our Own Backyard." A fifth essay explains the need to transform zoos and aquaria into

refuges for endangered species. Finally, "Life in the Next Millennium—
Who Will Be Left in Earth's Community?" argues for stewardship of the
earth and suggests ways to preserve it. Graphs, charts, and black-and-
white photographs are included.

Kennedy, Michael. **Australia's Endangered Species.** New York: Prentice
Hall Press, 1990. 192p. ISBN 0-13-053208-8.

A variety of conservation experts has contributed to this work, which
opens with an essay by Norman Myers, internationally known British
consultant and author of books on environment and development.
Chapters cover Australia's threatened mammals, birds, reptiles, amphibians, and fish. It is illustrated with color photographs by Australia's top
wildlife photographers. There are also extensive lists of Australian plant
and animal species that are threatened with extinction or presumed
extinct.

Norton, Bryan G., ed. **The Preservation of Species—The Value of Biological Diversity.** Princeton, NJ: Princeton University Press, 1986. 303p.
ISBN 0-691-08389-4.

As stated in the preface, the purpose of this book is "to analyze and
interpret existing scientific data, presenting it in conjunction with analy-
ses of the problem and its possible solutions." The book provides an
overview of the preservation problem and acts as a guide for legislators,
policymakers, and people not trained in biology. The focus is on understanding the factors that affect decision-making in the United States
and from this understanding, appreciating how international decisions
are made.

Schultes, Richard Evans, and Robert F. Raffauf. **The Healing Forest:
Medicinal and Toxic Plants of the Northwest Amazonia.** Portland, OR:
Dioscorides Press, 1990. 484p. ISBN 0-931146-14-3.

An ethnobiologist and a phytochemist are the authors of this work that
describes 1,516 plant species in the Northwest Amazonian rainforest.
Little or no chemical analysis has been conducted on many of these
species. Experts estimate that there could be as many as 80,000 plants yet
to be discovered and researched in the area. The authors call attention to
the need for conservation and emphasize the importance of ethnobiol-
ogy. It is their hope that this reference will alert others to the importance
of Amazonian plants and also further study of Indian folklore.

Stone, Roger D. **Wildlands and Human Needs: Reports from the Field.**
Baltimore, MD: World Wildlife Fund, 1991. 151p. ISBN 0-942635-17-52.

This compilation of reports describes World Wildlife Fund projects in
the Caribbean, Central and South America, Africa, and Southeast Asia.

The projects focus on sustainable management of wildlife areas and protection of the people who live in these regions. Although many of the projects carried out in conjunction with other environmental and human rights organizations have brought positive results, the ongoing efforts to preserve wildlife face huge barriers as the many inhabitants of the earth compete for resources.

Teitel, Martin. **Rain Forest in Your Kitchen: The Hidden Connection between Extinction and Your Supermarket.** Covelo, CA: Island Press, 1992. 120p. ISBN 1-55963-153-8.

As the title suggests, this book shows how the decisions Americans make in the supermarket can affect the way giant agribusinesses operate. Teitel discusses the loss of biodiversity in the sources of U.S. food supply and how agribusiness—large corporations—have narrowed food choices to mass-produced plants and animals, reducing the gene pool and threatening the genetic resources that may be needed in the future. He suggests ways that consumers can modify their eating and buying habits and help protect plant and animal species.

Terborgh, John. **Where Have All the Birds Gone?** Princeton, NJ: Princeton University Press, 1989. 207p. ISBN 0-691-02428-6.

In this book, Terborgh follows the migrating paths of 250 species of warblers, shorebirds, and ducks on their way to tropical countries in the southern hemisphere. He documents what has happened to the land where the birds rest and feed while migrating. His principal message is that if virgin forests are continually exploited and the excessive use of pesticides persists, many species of birds will disappear.

Wilson, Edward O., ed. **Biodiversity.** Washington, DC: National Academy Press, 1988. 521p. ISBN 0-309-03783-2.

Leading researchers on biodiversity and rainforest ecology and policymakers prepared the 57 papers included in this volume. The papers were the outcome of a 1986 public forum on BioDiversity under the auspices of the National Academy of Sciences and the Smithsonian Institution. In the words of the editor, a distinguished biologist, the book updates "many of the principal issues in conservation biology and resource management. It also documents a new alliance between scientific, governmental, and commercial forces . . . [which could] reshape the international conservation movement for decades to come."

Wilson, Edward O. **The Diversity of Life.** Cambridge, MA: Harvard University Press, 1992. 464p. ISBN 0-674-21298-3.

Called a "bible of the biosphere," this book guides the reader through evolutionary time, describing and showing with color photographs and illustrations the diversity of the biological world. Wilson, who is considered the dean of biodiversity studies, describes not only the threats that human activities pose to our biological wealth but also shows how the preservation of that wealth can be a part of economic development.

General Environmental Subjects

Brown, Lester R., ed. **The World Watch Reader on Global Environmental Issues.** New York: W.W. Norton, 1991. 336 p. ISBN 0-393-03007-5.

This is an anthology of articles that have appeared in *World Watch* magazine. The articles provide an in-depth coverage of how the earth is faring and a vision of how to create an environmentally responsible future. People concerned about air and water pollution, deforestation, and other environmental problems will find this anthology of interest.

Corson, Walter H., ed. **The Global Ecology Handbook: What You Can Do about the Environmental Crisis.** Boston: Beacon Press, 1990. 414p. ISBN 0-8070-8501-4.

Written by the staff of the nonprofit Global Tomorrow Coalition (see Chapter 7 for a description), this practical guide discusses the current state of the environment in relation to economic and social issues and energy policy. Chapters cover tropical rainforests, global warming, biological diversity, and many other environmental issues. The handbook shows how individuals and groups can take steps to protect the planet and includes tables, charts, and graphs as well as extensive chapter notes.

Darnay, Arsen J., ed. **Statistical Record of the Environment Worldwide.** New York: Gale Research, Inc., 1991. 855p. ISBN 0-8103-8374-8.

Except for introductory text, this book, as the title suggests, is a compilation of data and tables on environmental topics, with the focus primarily on North America. Statistics come from a variety of sources and should be of interest to those concerned about not only rainforests but many other environmental issues as well.

Eckholm, Erik. **Down to Earth.** New York: W.W. Norton, 1982. 238p. ISBN 0-393-01600-5.

Covering a number of global environmental issues, including chapters on "Deforesting and Reforesting the Earth" and "Biological Diversity and Economic Development," this book was prepared in commemoration of the tenth anniversary of the Stockholm Conference on the Human Environment. It stresses the links between the global poor and the fate of the world environment and shows how environmental protection and shared economic progress are compatible.

Ehrlich, Paul R., and Anne H. Ehrlich. **Healing the Planet.** New York: Addison-Wesley, 1991. 336p. ISBN 0-201-55046-6.

In this work, these two well-known scientists focus on the causes—the underlying problems rather than the symptoms—of environmental destruction. The authors recognize that the disparity between rich and poor nations must be addressed and that the problems of the southern hemisphere need to be resolved in order to make progress in terms of environmental protection. The book includes extensive source notes.

Gore, Albert. **Earth in the Balance: Ecology and the Human Spirit.** Boston: Houghton Mifflin, 1992. 407p. ISBN 0-395-57821-3.

As a U.S. senator from Tennessee and as vice president of the United States, Al Gore has worked for environmental protection. He explains in this book how he reached the conclusion that human civilization's "ravenous appetite for resources" threatens the ecological balance on earth. Gore argues that we all need to reshape our attitudes and ideas about our relationship with nature if we are to save the planet from ecological disaster. He describes a "Global Marshall Plan" with five strategic goals: stabilizing world population, developing environmentally appropriate technologies, changing the way people measure the economic impact of their decisions to include long-term costs to the environment, negotiating international agreements to protect the world's ecology, and educating people around the world on the need to conserve the earth's resources. Gore also points out the need for social and political justice so that sustainable societies can exist. Extensive chapter notes and bibliography are included.

Lovelock, James. **The Ages of Gaia: A Biography of Our Living Earth.** New York: W.W. Norton, 1988. 252p. ISBN 0-393-02583-7.

Lovelock first spelled out his Gaia theory, named for the Greek goddess of earth, in a 1979 best seller, *Gaia: A New Look at Life on Earth.* This book uses scientific data from several disciplines to support the theory that the earth is a living whole. Lovelock shows how human activities such as fossil fuel burning and deforestation that contribute to the greenhouse effect threaten the health of the planet and ultimately human life itself.

Lovelock, James. **Healing Gaia: Practical Medicine for the Planet.** New York: Harmony Books, 1991. 192p. ISBN 0-517-57848-4.

Building on his Gaia theory, Lovelock takes on the role of the "planetary physician" to diagnose and suggest cures for environmental ailments.

However, much of this book attacks the many critics of the controversial Gaia hypothesis—scientists who do not accept the concept that the earth should be seen as one living organism.

McKibben, Bill. **The End of Nature.** New York: Random House, 1989. 226p. ISBN 0-87701-814-6.

"Nature, we believe, takes forever." So begins McKibben's book, which shows that because of human activities, "our reassuring sense of a timeless future . . . is a delusion." McKibben argues that global warming and depletion of the ozone layer are already destroying nature as we know it and he makes a compelling plea for people to be caretakers of the earth and custodians of all life forms.

Global Warming

Oppenheimer, Michael, and Robert H. Boyle. **Dead Heat.** New York: Basic Books, 1990. 268p. ISBN 0-465-09804-5.

"Humanity is hurtling toward a precipice. Left unchecked, the emissions of various gases, particularly carbon dioxide from fossil-fuel combustion and deforestation, are likely to alter the Earth's climate so rapidly and so thoroughly as to destroy much of the natural world. . . . But such an outcome is not inevitable." So begins the prologue to this book, written by a senior scientist with the Environmental Defense Fund and a senior writer for *Sports Illustrated.* The opening statements signal the content of this book: an action plan for ways that industries and individuals can reverse the trend toward global warming. Includes source notes.

Revkin, Andrew. **Global Warming: Understanding the Forecast.** New York: Abbeville Press, 1992. 180p. ISBN 1-55859-310-1.

Illustrated with dramatic color and black-and-white photographs and a variety of drawings, this book was published to complement a major traveling exhibition by the same title prepared in conjunction with the American Museum of Natural History in New York City. Award-winning journalist Andrew Revkin explains how scientists study past and present climate to predict future changes and how human activities, such as burning fossil fuels and deforestation, lead to a buildup of carbon dioxide and other greenhouse gases. In a chapter called "Business as Usual," Revkin outlines the dire consequences of not taking action now to reduce CO_2 emissions. The final chapter, "A Greenhouse Diet," tells what individuals can do to reduce the risk of global warming. Included also are lists of organizations and resources and suggested reading.

Schneider, Stephen H. **Global Warming: Are We Entering the Greenhouse Century?** New York: Vintage Books/Random House. 343p. ISBN 0-679-73051-6 (paper).

Stephen Schneider is director of Interdisciplinary Climate Systems at the National Center for Atmospheric Research in Boulder, Colorado, and frequently testifies on environmental policy before U.S. Congressional committees and governments of other countries. His book describes in understandable terms the scientific research and data backing up the global warming theory and what global policies and strategies are needed to cope with the "greenhouse century" ahead.

Indigenous Peoples

Burger, Julian. **The Gaia Atlas of First Peoples: A Future for the Indigenous World.** New York: Gaia/Anchor, 1990. 191p. ISBN 0-385-26653-7.

Written in a clear and graphic style, this volume is a unique source book for anyone interested in the lives of first peoples, including indigenous people of the rainforests. Fifty concise essays, illustrated with photographs, maps, charts, and other graphics, describe native peoples worldwide and show how resource management, herbal medicine, cooperation, and conflict resolution are intrinsic to many indigenous cultures.

Clay, Jason W. **Indigenous Peoples and Tropical Forests: Models of Land Use and Management from Latin America.** Cambridge, MA: Cultural Survival, 1988. 116p. ISBN 0-939521-38-5.

How do indigenous peoples of the tropical rainforests use and sustain the resources of their regions? That is the basic question explored in this report by the former director of Cultural Survival. The report summarizes activities such as hunting and gathering and various types of agriculture that sustain indigenous populations and their environment. A bibliography of more than 400 works is included.

Denslow, Julie Sloan, and Christine Padoch. **People of the Tropical Rainforest.** Berkeley: University of California Press and Smithsonian Institution Traveling Exhibition Service, 1988. 231p. ISBN 0-520-06295-7 (cloth); 0-520-06351-1 (paper).

In the words of this book's authors, "The people of the tropical rainforests of the world, while all one species with similar needs, capabilities, and tolerances, are also an enormously varied lot." In a collection of essays, contributors to this work describe the cultures of such groups as the Kayapo, the Lacandon Maya, the Pygmies of the Congo basin, and the Hmong and Lua of Thailand, and more recent arrivals to the rainforest who depend on slash-and-burn agriculture for their staple crops and on the forest and rivers for game and fish. This book, as the preface notes, "offers no solutions to the 'rain forest problem,'" but contributors do explore the diverse ways that people use the forest; the widespread impact of such business ventures as logging, plantation forestry, mining, and cattle ranching; and some projections about the future of tropical rainforests and the people who live in them. Includes color photographs.

Good, Kenneth, and David Chanoff. **Into the Heart: One Man's Pursuit of Love and Knowledge among the Yanomama.** New York: Simon & Schuster, 1991. 352p. ISBN: 0-671-72874-1.

In 1975 Kenneth Good, an anthropology student, began working with the Yanomama Indians of the Amazon rainforest. He lived in Yanomama communities for 12 years, studying their way of life, learning their language, and eventually marrying a woman chosen for him by the tribe. Although Good describes the violence the Indians sometimes inflict upon one another, he also shows the respect and kindness that prevails. This book is not only a story of the culture of an indigenous people but also a personal account of Good's relationship with his wife, Yarima, whom the headman selected for him when she was 9 years old. Good waited for Yarima to come of age for marriage and fell in love with her. Because he was an outsider, he encountered problems with Venezuelan officials and nearly lost his wife, but he was able to take her with him when he returned to the United States to teach at Jersey City State College in New Jersey. The book is illustrated with photographs.

Guss, David M. **To Weave and Sing: Art, Symbol, and Narrative in the South American Rainforest.** Berkeley: University of California Press, 1989. 274p. ISBN 0-520-06427-5.

The Yekuana Indians of Venezuela's dense rainforest and their culture are the subject of this illustrated work. Guss concentrates on symbols used in basket weaving and asserts that by studying the history of the baskets, one can learn the history of the Yekuana culture and how the people have adjusted to Western ways. Other chapters cover such topics as architectural forms and the use of herbs and body paints.

Hansen, Eric. **Stranger in the Forest: On Foot across Borneo.** Boston: Houghton Mifflin, 1988. 286p. ISBN 0-395-44093-9.

In 1982, Eric Hansen began his 1500-mile walk through the Borneo rainforest, keeping copious notes of his adventures along the way. Not only is this a story of Hansen's journey but it is also a look at the Penan, indigenous people whose way of life is threatened by twentieth century "developments." A gifted storyteller, Hansen brings the rainforest to life with humor, compassion, and attention to detail.

Hemming, John. **Amazon Frontier: The Defeat of the Brazilian Indians.** Cambridge, MA: Harvard University Press, 1987. ISBN 0-674-01725-0.

Hemming begins his story in 1755 when the Brazilian tribes were liberated from the Portuguese government. This is the story of how the Indians' way of life was destroyed and how disease from the colonists decimated their tribes. Hemming divides the story into four parts: The Directorate; Independence; Amazonia: The Rubber Boom; and Missionaries, Anthropologists, and Indian Resistance.

Sustainable Society

Brown, Lester R., Christopher Flavin, and Sandra Postel. **Saving the Planet: How To Shape an Environmentally Sustainable Global Economy.** New York: W.W. Norton, 1991. 224p. ISBN 0-393-03070-9 (cloth); ISBN 0-393-30823-5 (paper).

The first in a World Watch Institute series on the environment, this volume strives to answer the most fundamental question of today's generation: How can we create a world economy and not destroy the earth in the process? *Saving the Planet* evokes a vision of a global economy that does not compromise the prospects of future generations. To achieve this vision, the authors offer a way to restructure energy and economic systems and aid distribution programs. Includes chapter notes.

Costanza, Robert, ed. **Ecological Economics: The Science and Management of Sustainability.** New York: Columbia University Press, 1991. 525p. ISBN 0-231-07562-6.

Forty-two experts contributed to this technical work for college students. The book, which is the result of the first biannual workshop of the International Society for Ecological Economics, links economics and ecology and shows how countries can develop economic policies that do not destroy global ecosystems. It is divided into three parts: "Developing an Ecological Economic World View," "Accounting, Modeling and Analysis," and "Institutional Changes and Case Studies." Contributors explore the research, training programs, and complex techniques needed to deal effectively with environmental problems.

Darmstadter, Joel, ed. **Global Development and the Environment: Perspectives on Sustainability.** Washington, DC: Resources for the Future, 1992. 91p. ISBN 0-915707-63-2.

Using an interdisciplinary approach and written for those who have some understanding of economics, ecology, and sustainability, this book is a compilation of essays addressing such topics as biodiversity, population control, the wise use of natural resources, and alternative energy sources. The volume provides an overview of sustainable development and includes practical solutions to environmental problems, along with actions that students and others can take to bring about sustainability.

Gupta, Avijit. **Ecology and Development in the Third World.** London and New York: Routledge, 1988. 80p. ISBN 0-415-00673-2.

This introductory look at development in nonindustrialized nations is designed for students of geography, the environment, or development issues. Gupta describes the problems created when nations develop their economies but ignore the ecology. The book covers the impacts of development on land and water resources and air quality and one chapter focuses on rainforests around the world, including discussions on the demand for forest resources and the effects of deforestation. Case studies and key ideas are part of each chapter. Black-and-white photographs, diagrams, maps, and charts are included.

Harrison, Paul. **The Third Revolution: Environment, Population and a Sustainable World.** New York: St. Martin's Press (U.S. distributor), I.B. Tauris (publisher), 1992. 351p. ISBN 1-85043-501-4.

In this call for a third revolution (going beyond the agricultural and industrial revolutions) in favor of sustainable development, Harrison explains how overconsumption, rapidly increasing population, and destructive technologies have contributed to global ecological problems. He describes from firsthand knowledge and field work in Asian countries such as Bangladesh the disasters brought on in part by overpopulation. Building on work by Paul Ehrlich, Harrison explains the formula Population $\times$ Consumption $\times$ Technology = Environmental Impact and shows with specific examples how population growth and environmental deterioration are linked. The book also includes examples of steps that can be taken to stabilize population, such as granting women rights equal to those of men and providing access to birth control and healthcare. In Harrison's view, the decisions made in the 1990s will determine whether or not the global third revolution will succeed.

Piasecki, Bruce, and Peter Asmus. **In Search of Environmental Excellence.** New York, London, and Tokyo: Simon & Schuster, 1990. 200p. ISBN 0-671-69090-6.

This book presents a variety of suggestions on how ordinary citizens, public officials, policymakers, activists, and industry leaders can work toward resolving environmental problems and includes ideas for controlling deforestation. The book is essentially a blueprint for survival and a call to action to save the planet.

Plotkin, Mark, and Lisa Famolare. **Sustainable Harvest and Marketing of Rain Forest Products.** Washington, DC: Island Press, 1992. 288p. ISBN 1-55963-169-4 (cloth); 1-55963-168-6 (paper).

In 1991, a Panama City Workshop brought together leaders of indigenous groups, conservationists, academic researchers, and business executives to share ideas on how to harvest nontimber products at sustainable levels and to discuss ways to bring products to international markets. This book is a compilation of workshop papers.

Repetto, Robert. **World Enough and Time.** New Haven: Yale University Press, 1986. 147p. ISBN 0-300-03648-5.

In this book, Repetto summarizes the Global Possible Conference sponsored by the World Resources Institute. He also suggests specific ways to use earth's resources wisely and provides examples of how low-cost policy changes can bring about improvements in the quality of life. The final chapter explores ways that major institutional groups can play a special role in instituting and developing policy for resource management.

Repetto, Robert, ed. **Public Policies and the Misuse of Forest Resources.** Cambridge, MA: Cambridge University Press, 1988. 432p. ISBN 0-521-34022-5 (cloth); 0-521-33574-4 (paper).

The World Resources Institute sponsored the research that led to this work, which is part of a broader examination of policy changes needed worldwide to bring about ecologically sound economic growth. Contributors to this book show how reasoned economic policies can help check rapid deforestation in Indonesia, Malaysia, the Philippines, China, Brazil, West Africa, and the United States. An important message the book conveys is: "more reasonable policies can save both natural and financial resources."

Roddick, Anita. **Body and Soul: Profits with Principles—the Amazing Success Story of Anita Roddick and The Body Shop.** New York: Crown Publishing Group, 1991. 256p. ISBN 0-517-58542-1.

This story of an environmental activist and businesswoman explains how Roddick made millions for her investors and yet did not sell her soul. Roddick has become one of the most outspoken activist businesswomen in the world. She developed The Body Shop, manufacturing cosmetics and skin care products from "natural" sources and competing in what she calls the "nastiest industry in the world." Roddick describes how her company espouses social and environmental responsibility and acts upon those principles through "Trade Not Aid" programs in the Amazon and other parts of the world.

Stone, Roger D. **The Nature of Development.** New York: Knopf, 1992. 304p. ISBN 0-394-58358-2.

Using evidence from various environmental organizations and development agencies, Stone shows how many international development efforts have failed to reduce the poverty of millions of people. As a result, the poor in developing nations have placed demands on global resources— for example, cutting trees on their land for meager profits or overfishing for food. Stone, of the World Wildlife Fund, also describes some successful grassroots initiatives to conserve resources in developing countries.

Wellner, Pamela, and Eugene Dickey. **The Wood Users Guide.** San Francisco: Rainforest Action Network, 1991. 64p. ISBN 0-9628033-16.

The authors wrote this guide to help further the campaign of the Rainforest Action Network to save temperate and tropical rainforests. Designed to help wood users identify rainforest timber and select alternatives, the guide gives comprehensive lists of temperate and tropical woods and descriptions of their color and use. Nonwood alternatives and where to find them are listed along with "Ecologically-Minded Lumber Suppliers."

Temperate Rainforests

Davis, John, ed. **The Earth First! Reader: Ten Years of Radical Environmental Journalism.** Layton, UT: Gibbs Smith Publishers, 1991. 240p. ISBN 0-87905-387-9.

Dave Foreman, cofounder of the Earth First! movement, wrote the foreword for this compilation of 40 feature articles from the discontinued *Earth First! Journal.* The articles provide a history of this radical environmental movement and the concepts behind it. Using confrontational tactics and on occasion illegal methods, the *Earth First!* movement gained widespread publicity during the 1980s and called attention to environmental destruction, particularly in the ancient forests of the Pacific Northwest. A final section explains why the movement split in 1990.

Dietrich, William. **The Final Forest: The Battle for the Last Great Trees of the Pacific Northwest.** New York: Simon & Schuster, 1992. 290p. ISBN 0-671-72967-5.

A Pulitzer Prize–winning correspondent for the *Seattle Times* is the author of this anecdotal and clearly presented account of the people affected by controversies over the ancient forests of the Pacific Northwest, particularly the Olympic National Forest. In each chapter, Dietrich presents stories of people from a broad range of backgrounds—ecologists, mill workers, biologists, industrialists, truckers, politicians, forest rangers, and others. Yet the book also shows that the forest "is perhaps a metaphor for all that we cherish and exploit on earth." In his final words, Dietrich makes his point: "This book is, of course, written on a tree. . . . I hope you read it for whatever understanding it provides. Then, when you get a chance, go and read the living things that it came from."

Ecotrust and Conservation International. **Coastal Temperate Rain Forests: Ecological Characteristics, Status and Distribution Worldwide.** Occasional Paper Series No. 1. Portland, OR: Ecotrust, 1992. 64p.

Ecotrust and Conservation International prepared this document to solicit comments from researchers and managers working in coastal temperate rainforests around the world. The report provides a provisional definition of coastal temperate rainforests and an overview of the subject. It also includes preliminary maps and tables illustrating the "original" extent of coastal temperate rainforests.

Ellis, Gerry, and Karen Kane. Foreword by Gaylord Nelson. **America's Rain Forest.** Minocqua, WI: NorthWord Press, 1991. 160p. ISBN 1-55971-129-9.

Gerry Ellis, world-renowned photographer and naturalist, and Karen Kane, author of many books and audiovisual scripts on nature are a husband-wife team who celebrate the ancient rainforests of the Pacific Northwest in this work. It covers the rainforests that stretch in an "emerald string" from northern California through British Columbia to southeast Alaska's Tongass forest, which is the largest tract of temperate rainforest in the world. Illustrated with 170 color photographs, the book looks at the geology, climate, and biological diversity of the forests and how they are interconnected and describes the people who once lived in the region. The authors call for action to protect America's threatened rainforests.

Ervin, Keith. **Fragile Majesty: Battle for North America's Last Great Forest.** Seattle: The Mountaineers, 1989. 272p. ISBN 0-8986-230-2.

Arguments on preserving the old-growth forests are the thrust of this book. The issues are complex and concern not only the northern spotted owl controversy but also the timber practices of the U.S. Forest Service, timber market exports, and reforestation practices. On the whole *Fragile Majesty* is a well balanced and thoughtful look at temperate rainforests in the Pacific Northwest.

Hartzell, Hal, Jr. **The Yew Tree—A Thousand Whispers.** Eugene, OR: Hulogosi, 1991. 319p. ISBN 0-938493-13-2.

According to the author, this "biography of a species" began as "a book about the myth, legend, lore, historical and poetical associations of the yew tree." But Hartzell, a treeplanter, rewrote the book to trace not only the history of the yew but to include botanical and geographical information about this ancient tree, said to be over 200 million years old. The book includes the current conflict over the only natural stands of yew remaining in the Pacific Northwest: how the discovery of taxol, a promising anti-cancer chemical occurring in yew bark, threatens to destroy the yew tree population. Hartzell documents the politics of the debate and the roles of the National Cancer Institute, the U.S. Forest Service, and chemical companies in attempts to exploit the yews of the U.S. temperate rainforest.

Herndon, Grace. **Cut and Run: Saying Goodbye to the Last Great Forests in the West.** Telluride, CO: Western Eye Press, 1991. 239p. ISBN 0-941283-11-9.

Written by an investigative newspaper reporter, this book defines the issues and chronicles the current forest and logging controversies in the western United States. The focus is on U.S. Forest Service timber policies and how those policies have affected the West. The author includes personal interviews with people of each state.

Kelly, David, and Gary Braasch. **Secrets of the Old Growth Forest.** Layton, UT: Gibbs Smith Publishers, 1988. 99p. ISBN 0-87905-174-4.

Without delving into the maze of issues concerning old-growth forests, the authors have produced an informative and stunning look at the ancient forests of the Pacific Northwest. In a large format, the book's lush photographs help the viewer gain an appreciation for ancient forests. An informative appendix called "One Forest, Many Battlegrounds" describes the various parcels of ancient forestland.

Ketchum, Robert, and Carey D. Ketchum. **The Tongass: Alaska's Vanishing Rain Forest.** New York: Aperture, 1987. 112p. ISBN 0-89381-266-8.

This large-format book contains beautiful photographs by a landscape photographer of a little-known rainforest, the Tongass in Alaska. Brief text provides an introduction to the various threats to the Tongass. The book is not meant to be a study of the rainforest, but rather it is designed to foster awareness of and appreciation for its subject.

Kirk, Ruth, with Jerry Franklin. **The Olympic Rain Forest: An Ecological Web.** Seattle, WA: University of Washington Press, 1992. 128p. ISBN 0-295-97195-9 (cloth); ISBN 0-295-97187-8 (paper).

Author and photographer Ruth Kirk lived in the Olympic National Forest while her husband served as a ranger-naturalist with the Olympic

National Park. She coauthored this book with Jerry Franklin, who was chief plant ecologist for the U.S. Forest Service and now is Professor of Ecosystem Analysis at the University of Washington. Together the two have prepared a readable and scientifically sound text that focuses on the Olympic National Forest on Washington's Olympic Peninsula and its unique ecological web. The book includes line drawings, maps, and 100 color photographs that capture the variety and grandeur of this magnificent forest, which has been recognized as a National Park, a World Biosphere Reserve, and a World Heritage Site.

Maser, Chris. **Forest Primeval: The Natural History of an Ancient Forest.** San Francisco, CA: Sierra Club Books, 1989. 282p. ISBN 0-87156-683-4.

Maser begins in the year 988 and traces the history of an ancient forest in the Cascade Mountain range in Oregon. He presents a historical timeline of events that occurred during the lifetime of the mountain trees. Along the way, he presents the natural history and biology of the flora and fauna of the forest. This volume is a unique comparison of the life of ancient fir trees to that of the human history taking place. It includes references, a glossary, and black-and-white photographs.

Middleton, David. **Ancient Forests.** San Francisco: Chronicle Books, 1992. 107p. ISBN 0-87701-814-6.

With beautiful photographs and text, Middleton defines an old-growth forest and some of its inhabitants. Although he attempts to present answers to the question about whether old-growth forests should be cut, he does not address the complex issues associated with logging practices in the Pacific Northwest. However, the book accomplishes its purposes: to enhance appreciation for old-growth forests.

Morrison, Peter H., and The Wilderness Society. **Old Growth in the Pacific Northwest: A Status Report.** Washington, DC: The Wilderness Society, November 1988. 46p.

One of a number of reports and policy papers on national forests produced by The Wilderness Society, this report is the result of a study conducted by Peter Morrison, an experienced forest ecologist. According to The Wilderness Society, the report "provides for the first time a valid, scientific assessment of the amount and condition of ecological old growth that still exists on 6 of the 12 westside national forests"—those on the western side of the Cascade Mountains in Washington, Oregon, and Northern California. Illustrated with charts, tables, graphs, and maps, the report shows that Morrison's findings conflict with higher inventories of old growth reported by the U.S. Forest Service. However, a Wilderness Society analysis of USFS inventories indicates that the Forest

Service used some inaccurate data and outdated information to reach its conclusions.

Norse, Elliot. **Ancient Forests of the Pacific Northwest.** Covelo, CA: Island Press, 1990. 327p. ISBN 1-559-63017-5.

Norse's book describes ancient forests and their locations and climate and presents a history of forest use by humans; it also discusses the forest ecosystem in detail and introduces the concept of biological diversity. However, the book is essentially an analysis of the U.S. Forest Service's long-term management plans, concluding that the plans are inadequate due to lack of information and do not allow for reasoned decisions.

O'Toole, Randal. **Reforming the Forest Service.** Covelo, CA: Island Press, 1988. 248p. ISBN 0-933280-49-1 (cloth); ISBN 0-933280-45-9 (paper).

Randal O'Toole has been called "the Adam Smith of forest economics." A forest economist, O'Toole based his book on years of research, reviews of dozens of U.S. Forest Service (USFS) plans and thousands of timber sales, and a thorough study of the USFS. O'Toole proposes several basic changes in the USFS, such as funding the service with net receipts from timber rather than appropriations from Congress, removing incentives to sell timber below cost, and charging recreation fees for use of wilderness areas. He also includes a proposal for natural resource protection.

Robinson, Gordon. **The Forest and the Trees.** Covelo, CA: Island Press, 1988. 257p. ISBN 0-933280-40-8 (paper).

Forestry professionals and laypersons concerned about the environment will find this book essential reading. A respected forester with 50 years' experience, Robinson provides a detailed look at the way forests in the United States are managed and frequently mismanaged today. The book includes a history of national forest management and descriptions of management practices that protect forest life while providing sustained yields. In the last section, the author presents summaries of published research and opinions on forest management.

Zuckerman, Seth. **Saving Our Ancient Forests.** Los Angeles: Living Planet Press, 1991. 116p. ISBN 0-9626072-9-0.

Based on research sponsored by The Wilderness Society, this is a story of the ancient forests in the Pacific Northwest written in a concise and lively manner. It covers such issues as what ancient forests are, the ecology of the forests, how they are threatened by logging, and what can be done about the problem of disappearing forests. The book is illustrated and contains numerous sidebars with "fun facts" about forests.

Tropical Rainforests

Almeda, Frank, and Catherine M. Pringle, eds. **Tropical Rainforests: Diversity and Conservation.** San Francisco, CA: California Academy of Sciences, 1988. 206p. ISBN 0-940228-19-X.

A review of research and conservation efforts in Costa Rica, this volume is a compilation of papers presented at the California Academy of Sciences during a 1985 symposium on diversity and conservation of rainforests. Two-thirds of the book is devoted to the experiences of the Organization of Tropical Studies and its research in Costa Rica, including a good history of the organization and a look at the status of Zona Protectora in Costa Rica.

Anderson, Robert S., and Walter Huber. **The Hour of the Fox: Tropical Forests, the World Bank, and Indigenous People in Central India.** Seattle: University of Washington Press, 1988. 158p. ISBN 0-295-96603-3.

A communications specialist and an anthropologist look at a proposed rainforest project in India. With funding from the World Bank, the project was designed to convert India's largest remaining tropical forest to a plantation. The authors discuss the developmental and environmental conflicts within the context of fundamental policy failures, but nearly the entire book concerns the plight of the indigenous peoples of the forest. The project was canceled before the book was published.

Beazley, Mitchell, ed. **The International Book of the Forest.** New York: Simon and Schuster, 1981. 224p. ISBN 0-671-41004-0.

In the foreword, this book is called "an enthralling journey of exploration through the forests and jungles of the six continents . . . the first to examine the beauty, uniqueness, and crucial importance of *all* the world's forests." Beginning with a discussion of how the forest ecosystem works and people's relationships with the forest, the book covers diverse types of forests in Asia, Africa, Australia, Europe, and North and South America. Subsections deal with the Olympic Rainforest in the United States and various tropical rainforests. The book includes excellent photographs, drawings, and maps.

Borota, Ján. **Tropical Forests: Some African and Asian Case Studies of Composition and Structure.** New York: Elsevier, 1991. 274p. ISBN 0-444-98768-1.

Intended for scientists, foresters, college instructors, and forestry students, this work is a synthesis of Barota's studies on forest ecosystems. It evaluates scientific literature on tropical forests and provides basic scientific information to assist those planning rational and ecologically sound use of forests. The author is particularly detailed in his studies of forests in Ghana, Congo, Gabon, Tanzania, and Laos. The book includes maps and photographs.

Caufield, Catherine. **In the Rainforest: Report from a Strange, Beautiful, Imperiled World.** Chicago: University of Chicago Press, 1986. 306p. ISBN 0-226-09786-2.

Along with Caufield's *In the Rainforest* published by Knopf in 1984, this book has been read and studied extensively by those involved in rainforest preservation. It describes rainforest regions worldwide and explains the political and economic causes for rainforest destruction.

Cowell, Adrian. **The Decade of Destruction: The Crusade To Save the Amazon Rain Forest.** New York: Holt, 1990. 215p. ISBN 0-8050-1494-2.

Written by a British filmmaker who has documented the destruction of the Amazon rainforest since the 1950s, this book covers such topics as the International Monetary Fund, the debt crisis, and ranching interests—all of which have contributed to loss of rainforests. Cowell also describes how the warring tribes of the Amazon are joining forces to stop the devastation of their homelands. He offers hope that a combination of international and internal pressures will eventually halt the destruction.

Dalton, Stephen, George Bernard, and Andrew Mitchell. **Vanishing Paradise: The Tropical Rainforest.** Woodstock, New York: The Overlook Press, 1990. 176p. ISBN 0-87951-406-X.

This large-format book presents the work of Dalton and Bernard, nature photographers who are known worldwide. Many of the book's 200 dramatic and stunning photographs show tropical forest animals, birds, and plants in exquisite detail. Some photographs were taken in Costa Rica's rainforest. Zoologist Mitchell's text describes the methods by which a tropical rainforest is able to regulate moisture in the air, the variety of life in the forest, and how animals find food, hunt prey, and reproduce. A chapter on communication shows how mammals and birds stay in touch in a rainforest by being "very noisy, very colourful, or very smelly." An especially interesting section shows how mammals, birds, and insects are able to camouflage themselves. Other sections explain the forest canopy and how decay and renewal work hand-in-hand so that the forest can recycle itself.

Dixon, Anthony, Hannah Roditi, and Lee Silverman. **From Forest to Market: A Feasibility Study of the Development of Selected Non-Timber Forest Products from Borneo for the U.S. Market.** Cambridge, MA: Cultural Survival, 1991. (2 vols.) 300p. ISBN 0-962-9047-0-8.

Working with Cultural Survival (see Chapter 5), three Harvard Business School students carried out the studies that led to this report on how a business framework can be used to develop and market nontimber forest products from the rainforests of Borneo. Written as handbooks, the two volumes contain descriptions of products and their uses and how they can be developed, information on trade regulations and how to influence policymakers, and lists of individuals and groups working for business alternatives to forest destruction. Color photographs of 90 products are included.

Downing, Theodore E., Susanna B. Hecht, Henry A. Pearson, and Carmen Garcia-Downing, eds. **Development or Destruction: The Conversion of Tropical Forest to Pasture in Latin America.** Boulder, CO: Westview Press, 1992. ISBN 0-8133-7824-9.

Introducing this work, the editors point out that "The conversion of forest to grasslands in the humid tropics is one of the most profound land transformations of the 20th Century, with major consequences for biodiversity, global and regional climates, soil resources, and local populations." Contributors in the first section explain the livestock economy and how it relates to forest destruction. In the following sections, authors explain the environmental and social impacts of raising livestock in rainforest regions and alternatives to livestock production. Also included are the views of community representatives, peasants, and rubber tappers.

Dwyer, Augusta. **Into the Amazon: The Struggle for the Rain Forest.** San Francisco: Sierra Club Books, 1990. 250p. ISBN 0-87156-637-0.

In this wide-ranging book on Amazonia, Dwyer discusses the politics of the region, the struggles of Chico Mendes, and the myths and legends of the land. The author presents a passionate account of a vanishing way of life in Amazonia, describing how government, industrialists, and international exploiters have destroyed or are seeking to destroy an area of the world that is essential to the planet's survival.

Eden, Michael J. **Ecology and Land Management in Amazonia.** New York: Belhaven Press, 1990. 269p. ISBN 1-85293-118-3.

Management of renewable resources is the focus of this volume. Eden examines the major ecosystems in Amazonia: the forests, savannas, and wetlands. He also discusses Amerindian exploitation and methods for land use that do not destroy ecosystems and argues for more national parks in the region. This technical book includes an extensive bibliography and maps, charts, and graphs.

Fearnside, Philip M. **Human Carrying Capacity of Brazilian Rainforests.** New York: Columbia University Press, 1986. 291p. ISBN 0-231-06104-8.

Using scientific population models, Fearnside argues in this technical work that defining the density of population that can be sustained on an acceptable basis is one way to avoid degradation of the environment and ecosystems of the Brazilian rainforests. The book includes tables, a glossary, and an extensive bibliography.

Forsyth, Adrian, Michael Fogden, and Patricia Fogden. **Portraits of the Rainforest.** Ontario, Canada: Camden House, 1990. 156p. ISBN 0-921820-13-5.

A biologist and natural-history writer, Forsyth presents a collection of essays on diverse aspects of the tropical rainforest in this large-format book. Chapters describe the origins of varied plants and animals; why rare species are common in the tropics; and the nutrient cycles and roles of diverse life forms from wood-eating termites to rainforest amphibians to the "ultimate predator," the jaguar. Color photographs by the Fogdens enhance the text throughout.

Gallant, Roy A. **Earth's Vanishing Forest.** New York: Macmillan, 1991. 162p. ISBN 0-02-735774-0.

Written for young adults, this introductory book on the importance of the rainforest describes the forest's biodiversity and the threat to the homelands of indigenous peoples. Gallant also explains how the destruction of forests leads to loss of medicinal plants and contributes to the buildup of "greenhouse gases" in the atmosphere. He concludes that time is running out and the secrets of the rainforest could be lost forever. The book is illustrated with black-and-white photographs and includes maps and a glossary.

Gentry, Alwyn H. **Four Neotropical Rainforests.** New Haven, CT: Yale University Press, 1990. 627p. ISBN 0-300-04722-3.

This collection of papers was the result of a symposium on neotropical rainforests held in 1987. All of the papers are based on a comparison of ecosystem dynamics at four research stations: LaSelva in Costa Rica, Barro Colorado Island in Panama, Cocha Cashu in Peru, and Manaus in Brazil. The collection includes an extended section on forest dynamics.

Goodland, Robert, ed. **Race To Save the Tropics: Ecology and Economics for a Sustainable Future.** Washington, DC: Island Press, 1990. 219p. ISBN 1-55963-039-6 (cloth); 1-55963-040-X (paper).

The authors of the varied chapters in this book are what the editor calls "muddy-footed practitioners" who are committed to the concept that ecologically sound development makes economic sense and can become a reality. Along with solid scientific data, the authors provide practical examples of how economic ecology can be applied to forestry.

Gradwohl, Judith, and Russell Greenberg. **Saving the Tropical Forests.** Washington, DC: Island Press, 1988. 214p. ISBN 0-933280-81-5.

Released in conjunction with a major Smithsonian traveling exhibition, this book examines the deforestation of tropical rainforests and offers immediate and timely solutions in the form of case studies. The case studies are based on projects that have been implemented throughout the developing world and are short and informative—meant to be readable, not a thorough analysis. Source notes are included.

Griffiths, Michael. **Indonesian Eden: Aceh's Rainforest.** Baton Rouge, LA: Louisiana State University Press, 1990. 111p. ISBN 0-8071-1615-7.

Griffiths, who spent many years traveling through the Indonesian rainforests, presents his travels in journal form with lush color photographs. This large-format book includes photographs of rare animals and interviews with people who live in the rainforest.

Head, Suzanne, and Robert Heinzman, eds. **Lessons of the Rainforest.** San Francisco, CA: Sierra Club Books, 1990. 256p. ISBN 0-87156-682-6 (paper).

Essays from 24 leading authorities on the effects of worldwide destruction of tropical rainforests are included in this challenging book. The essays show how, in a global context, international economic and social patterns have affected the well-being of rainforests. Topics include "Tropical Forests and Life on Earth," "Five Hundred Years of Tropical Forest Exploitation," "Asia's Forest Cultures," "Extinction," and "Saving the Forest and Ourselves." The essay authors warn that a variety of reforms are needed to "ensure the sustainability of life on our planet" and they suggest ways to reach that goal.

Hecht, Susanna, and Alexander Cockburn. **The Fate of the Forest: Developers, Destroyers and Defenders of the Amazon.** London and New York: Verso, 1989. 266p. ISBN 0-86091-261-2.

This book traces the history of the Amazon rainforest over four centuries, explaining how explorers, conquerors, naturalists, and entrepreneurs viewed the riches of the Amazon. The study describes the realm of nature, the "heritage of fire" in the forest, exploitation of resources, and the many ways the region is being devastated through destruction of flora and fauna, poisoning of rivers, and persecution and killing of rubber tappers and settlers. A major section of the work covers the events leading up to the 1988 murder of Chico Mendes, leader of the rubber tappers. A Forest People's Manifesto with policies for agrarian reform and a call for an end to the oppression of forest peoples is appended. As the authors of the book point out, the rainforest will not flourish unless the people are able to prosper, too. There is an extensive bibliography and a map showing Amazon reserves as of early 1989.

Holm-Nielsen, L. B., L. C. Nielsen, and H. Baslev, eds. **Tropical Forests.** New York: Academic Press, 1989. 374p. ISBN 0-12-323550-6.

This book is a collection of technical and scientific papers that were presented at a symposium held at Aarhus University in Denmark. At the symposium, titled "Tropical Forests: Dynamics, Speciation and Diversity," experts presented studies conducted in Sri Lanka, Guiana, Borneo, China, and Amazonia. The editors for this work note in the preface that a baseline study needs to be completed on taxonomies and species before any new understanding of the tropical ecosystem can be articulated. The book includes charts, graphs, and an index of plant names.

Hurst, Philip. **Rainforest Politics: Ecological Destruction in Southeast Asia.** Atlantic Highlands, NJ: Humanities Press International, 1990. 320p. ISBN 0-86232-838-1 (cloth); 0-86323-839-X (paper).

This well-documented book describes the destruction of rainforests in Southeast Asian countries: Thailand, Indonesia, Malaysia, and the Philippines. Hurst explains the political and social aspects of deforestation and the conflict between the industrialized nations and nonindustrialized countries.

Jacobs, M., and R. A. A. Oldeman. R. Kruk, translator. **The Tropical Rain Forest: A First Encounter.** New York: Springer-Verlag, 1988. 311p. ISBN 0-387-17996-8.

This book is a thorough overview of the biological and human aspects of tropical rainforests. It covers how public awareness of rainforest issues has developed and how rainforests are studied. Other chapters discuss such topics as rainforest climate, soil, plants, and animals. Several chapters focus on the rainforests of specific areas (tropical Africa, for one). Other topics covered are the values of rainforests, damage and destruction, and efforts to protect these ecosystems. The book includes 170 illustrations.

Johnson, Brian. **Responding to Tropical Deforestation.** Baltimore, MD: WWF Publications, 1991. 63p. ISBN 0-89164-122-X.

This report is part of a series of research papers published by the World Wildlife Fund under the direction of the Osborn Center. The report contains an overview of the causes and consequences of the tropical rainforest crisis and analyzes three treaty negotiations and other agreements promoted by the United Nations conferences on Environment and Development. Johnson proposes to make maximum use of existing institutions rather than develop new ones.

Jordan, Carl F. **Nutrient Cycling in Tropical Forest Ecosystems—Principles and Their Application in Management and Conservation.** New York: John Wiley and Sons, 1985. 180p. ISBN 0-471-90449-X.

As the title suggests, this book is a technical work aimed at foresters and forestry students. Chapters cover "Factors which control nutrient cycles," "Nutrient conserving mechanisms," "Differences in ecosystem characteristics along environmental gradients," "Characterization of nutrient cycles," "Changes in nutrient cycles due to disturbance," and a summary of recommendations for managing tropical ecosystems. The book includes charts, graphs, and an extensive bibliography.

Kane, Joe. **Running the Amazon.** New York: Alfred A. Knopf, 1989. 227p. ISBN 0-394-55331-4.

In this true-life account, Joe Kane describes the first expedition ever to travel the full length of the Amazon River. Kane, who was the only American in the multinational crew, tells the story of how the crew learned to live together and to accept the challenges of the white-waters of the Amazon during the six-month journey. He also provides detailed descriptions of the land and river as well as the adventures of the expedition.

Kimerling, Judith, with the Natural Resources Defense Council. **Amazon Crude.** Washington, DC: The Natural Resources Defense Council, 1991. 132p. ISBN 0-9609358-3-5 (paper).

The devastating effects of virtually uncontrolled oil exploration in the Ecuadorean rainforests are the subject of this well-documented book. Not only are forests destroyed, but rainforest people are poisoned, debilitated, and killed by toxic pollutants from oil drilling and diseases introduced by outsiders. The book includes full-color photographs, maps, and illustrations.

Kricher, John. **A Neotropical Companion: An Introduction to the Animals, Plants and Ecosystems of the New World Tropics.** Princeton, New Jersey: Princeton University Press, 1990. 436p. ISBN 0-691-08520-X.

Written for the layperson, this overview of neotropical forests would be of interest to armchair travelers. Kricher defines rainforests and explains their functions and ecosystems. He briefly describes the medicinal plants and the many animals, birds, and reptiles found in the forests. A glossary is included.

Lamb, F. Bruce. **Rio Tigre and Beyond: The Amazon Jungle Medicine of Manuel Cordova.** Berkeley, CA: North Atlantic Books, 1985. 227p. ISBN 0-938190-61-X.

Picking up from his earlier book *Wizard of the Upper Amazon,* Lamb in this book completes the life story of Manuel Cordova, "a true bi-cultural healer," as he is called. A Peruvian rubber worker, Cordova was kidnapped by indigenous people of the Amazon and trained for seven years to be their new shaman, learning how to use plants as medicinals and to develop psychic powers, which he used extensively for diagnostic purposes. Anecdotal in style, this book is also a story of the healing properties of medicinal plants that are still in use today.

Lewis, Scott. **Rainforest Book.** Venice, CA: Living Planet Press, 1990. 112p. ISBN 0-9626072-1-5.

Dense with facts about rainforest animal and plant species, this book is written in layperson's terms. The author clearly explains why species are threatened and what can be done to save them. The last two chapters list resources and organizations working to prevent deforestation in the tropics.

Lieth, H., and M. J. Werger, eds. **Tropical Rainforest Ecosystems: Biogeographical and Ecological Studies.** (Ecosystems of the World 14B.) Amsterdam/New York: Elsevier, 1989. 713p. ISBN 0-444-42755-4.

A good general reference on the structure and function of rainforest ecosystems, this textbook is a major effort to compile current reference material on biogeographic and ecological aspects of the world's rainforests. Some of the most preeminent researchers are represented here. There is a discussion of general forest types along with current information on rainforest animals and plants. Concluding chapters discuss exploitation of Southeast Asian rainforests and Third World policies on forest conservation.

Longman, Kenneth A., and Jan Jenik. **Tropical Forest and Its Environment.** White Plains, NY: Longman Publishing Group, 1987. 347p. ISBN 0-582-44678-3.

Written by two European ecologists, this scientific and technical work is aimed primarily at students of ecology, whole-plant physiology, forestry, and natural resources. It is also a useful handbook for foresters, botanists, natural resource planners, and conservationists. According to the authors, "The aim of this book is to summarize available biological information on the nature of tropical forests and how they function, and to indicate some of the practical implications for anyone using or managing tropical forest land." The book includes extensive references and a species index along with a general index.

McIntyre, Loren. **Amazonia.** San Francisco: Sierra Club Books, 1991. 164p. ISBN 0-87156-641-9.

This large-format book is a collection of 121 lush color photographs by McIntyre, a photojournalist who has spent years photographing South America's wilderness. The structure of the book and titles of sections correspond to parts of the Amazon river. Photographs show the Amazon Basin and the "white water" of the mountainous region in the west; the "black water" of the Rio Negro in the north; and the "brown water" and "blue water" sections to the south and east respectively. Each section includes photographs of plant and animal life and human culture along with the author's insights in text form interwoven throughout.

Margolis, Mac. **The Last New World: The Conquest of the Amazon Frontier.** New York: W.W. Norton, 1992. 367p. ISBN 0-393-03379-1.

Although this is a story about the waste and ruin that has been the result of South America's version of Manifest Destiny, it is also the story of peasants, cattle ranchers, and rubber tappers who are trying to find ways to develop land areas without destroying complex ecosystems. Margolis shows that with the help of scientists and extension workers, people in the Amazon region are trying to find a middle course—between total preservation and total destruction. The book is illustrated with black-and-white photographs.

Martin, Claude. **Tropical Rainforests of West Africa.** Basel, Switzerland: Birkhauser Verlag; Boston: Birkhauser Boston, 1991. 235p. ISBN 0-8176-2380-9.

During the 1980s, approximately 7,200 square kilometers of rainforest were destroyed each year in West Africa. Arguing that the rainforests in West Africa are the forebears of what has happened and will happen to rainforests throughout the world, Martin discusses the history, ecology, conservation, and future of West Africa's rainforests. One of Martin's main contentions is that scientists have failed to set research priorities and to explore tropical forest complexities. This book is one of the few available on West African rainforests, and it includes descriptions of many species of primates and small mammals that live in the forests. It also explains conservation measures that are being taken. The book includes color photographs, charts, graphs, maps, and an extensive bibliography.

Miller, Kenton, and Laura Tangley. Introduction by Gus Speth. **Trees of Life: Saving Tropical Forests and Their Biological Wealth.** Boston: Beacon Press, 1991. 228p. ISBN 0-8070-8505-4.

Written for scientist and layperson alike, this book is part of the World Resources Institute Guide to the Environment series. The authors compare the historical treatment of North American forests to what is occurring now in Amazonia and focus on efforts to prevent deforestation. The book includes a chapter on old-growth forests and seven pages of forest facts.

Mitchell, Andrew. **The Enchanted Canopy: A Journey of Discovery to the Last Unexplored Frontier, the Rainforests of the World.** New York: Macmillan, 1986. 288p. ISBN 0-02-585420-8.

Written for a general audience as well as the informed layperson, this book includes 98 color photographs of tropical rainforest scenes. With interwoven text based on research conducted in rainforest regions worldwide, a major portion of the book explains and shows tree-top ecosystems, including photographs of insect life and predator-prey relationships unseen from the forest floor. The book also provides information on the people who live in close association with the rainforest canopy.

Mitchell, George J. **World on Fire: Saving an Endangered Earth.** New York: Macmillan, 1991. 320p. ISBN 0-684-19231-4.

One of the leading environmentalists in the U.S. Congress, Senate Majority Leader Mitchell examines global ecological crises, including climatic changes due to the greenhouse effect, overpopulation, and deforestation. He also describes the steps that must be taken to prevent major catastrophes. Mitchell explores the U.S. role in developing global conservation strategies, such as saving energy and shifting funds from military spending to environmental preservation.

Myers, Norman. **The Primary Source: Tropical Forests and Our Future.** New York: W.W. Norton, 1984. 399p. ISBN 0-393-01795-8.

A well-known author and environmental consultant, Myers clearly explains how tropical forests have been destroyed because of the profit incentives of multinational companies, the severe economic needs of Third World nations, and the overconsumption of developed nations. In this work, Myers offers a blueprint for solving some of the problems. The book includes references, tables, and black-and-white photographs.

Nations, James D. **Tropical Rainforests: Endangered Environments.** New York: Franklin Watts, 1988. 144p. ISBN 0-531-10604-7.

Written for young adults, this book stresses the need for immediate preservation of tropical rainforests. The author explains the importance of the forest ecology and biodiversity to people worldwide and discusses how some deforestation problems are being resolved. Illustrated with black-and-white photographs, the book also includes a current bibliography and a list of conservation organizations working to protect tropical rainforests.

Newman, Arnold. **Tropical Rainforest.** New York: Facts on File, 1990. 256p. ISBN 0-8160-1944-4.

A conservationist and science writer who has spent many years traveling to rainforests and studying their ecology, Newman brings together much detailed information in this work. It is divided into five sections: "What Is a Tropical Rainforest?" "The Web of Life," "Threats to the Forest," "What Do We Lose?" and "A Blueprint for Survival." Illustrated with photographs, this book is a comprehensive look at tropical rainforests and includes helpful appendixes listing educational resources, tropical timbers and domestic alternatives, and an essay titled "Toward Sustained Productivity." The extensive bibliography lists several hundred entries.

Nichol, John. **The Mighty Rain Forest.** London: David & Charles, 1990. 200p. ISBN 0-7153-9461-4.

Written in association with Worldforest 90, a week of international media events calling attention to rainforests of the world, this large-format book contains color and black-and-white photographs, many by nature photographer Adrian Warren. The text describes in an informal and sometime humorous style the rainforests of the world; the animals, plants, and people of the rainforests; and why the forests are in danger and what is being done to prevent destruction. Sidebars list a variety of statistics and interesting factual briefs about rainforests. The book also includes a list of organizations in various countries around the world working on rainforest conservation.

Payne, Junaidi. Photographs by Gerald Cubitt. **Wild Malaysia: The Wildlife and Scenery of Peninsular Malaysia, Sarawak and Sabah.** Cambridge, MA: MIT Press, 1990. 208p. ISBN 0-262-16078-1.

The rainforests of Malaysia in Southeast Asia were undisturbed just a century ago, but today the 13-state federation supplies more tropical wood products than any other region in the world. With more than 400 color photographs by a leading natural history photographer, this book shows the flora and fauna of Malaysia. Payne, who is project director of World Wide Fund Malaysia, describes the history and culture of Malaysia's people and how they are trying to prevent degradation of their environment.

Perlin, John. **A Forestry Journey: The Role of Wood in the Development of Civilization.** New York: W.W. Norton, 1989. 445p. ISBN 0-393-02667-1.

In this book, Perlin traces the role of wood in society, from the Bronze age to the nineteenth century, showing the role forests have played in the development of past civilizations. He explains that because wood was the primary fuel and building material in the past, its abundance or scarcity shaped the culture, demographics, economy, politics, and technology of each society. By examining how other societies dealt with the consequences of deforestation, Perlin provides insights that help people today understand and resolve problems associated with forest destruction.

Perry, Donald. **Life above the Jungle Floor.** New York: Simon & Schuster, 1986. 170p. ISBN 0-671-54454-3.

Two-thirds of animal and plant life in the rainforest resides in the canopy. In this volume, Perry, a pioneer in the exploration of rainforest canopies, describes the rainforest canopy of Costa Rica. He also explains how he devised a working area with a web of ropes and platforms from which he could study the biome. Color photographs illustrate the book.

Peterson, Dale. **The Deluge and the Ark: A Journey into Primate Worlds.** Boston: Houghton Mifflin, 1989. 378p. ISBN 0-395-51039-2.

Peterson's travels to rainforests around the world resulted in a story of his adventures and a look at 12 species and subspecies of primates and their habitats. Peterson explains why these animals are endangered and what is being done to save them. In his view, subsistence hunting and international trade of live primates are the deluge leading to extinction. The "ark" includes such protective measures as national parks and zoos.

Prance, Ghillean T., and Thomas E. Lovejoy. **Amazonia.** Elmsford, NY: Pergamon Press, 1985. 442p. ISBN 0-08-030-776-0.

This broadly focused book presents a comprehensive review of existing knowledge on the Amazon based on research of scientists actually studying in the area. It is divided into sections on physical setting, biology, and human impact, with chapters such as "Soils of the Amazon Rainforests," "Ecology of Amazon Primates," and "Agriculture in Amazonia." The volume presents the best of what is known about Amazonia and includes maps, diagrams, and black-and-white photographs.

Revkin, Andrew. **The Burning Season: The Murder of Chico Mendes and the Fight for the Amazon Rain Forest.** Boston: Houghton Mifflin, 1990. 298p. ISBN 0-395-52394-X.

This book is the story of Chico Mendes, a rubber tapper and environmental activist in Brazil who was murdered by wealthy cattle ranchers intent on clearing the forest and using the land for cattle grazing. The murder of Mendes brought cries of outrage and protest from around the world, and Mendes's name now symbolizes global efforts to save rainforests. Revkin describes events that led up to the murder, focusing on how Mendes and his fellow rubber tappers organized to preserve the forest upon which they depended for survival. He also describes the natural alliance between the tappers and the indigenous tribal groups who need the rainforest to live.

Ricciardi, Mirella. **Vanishing Amazon.** New York: Abrams, 1991. 240p. ISBN 0-8109-3915-0.

A renowned photographer, Ricciardi presents a photographic diary of her trip to Western Brazilian Amazonia and her observations of three Amazon Indian groups. In this work, which includes 220 photographs, she shows the daily life and culture of the Kempa, the Marubo, and the Yanomami, and explains how Amazon Indian groups have joined together to fight for preservation of their forest homelands. The text is brief, but anthropologist Marcus Colchester provides background information on each of the tribes.

Rifkin, Jeremy. **Beyond Beef: The Rise and Fall of the Cattle Culture.** New York: Dutton, 1992. 353p. ISBN 0-525-93420-0.

"There are currently 1.28 billion cattle populating the earth. They take up nearly 24 percent of the landmass of the planet and consume enough grain to feed hundreds of millions of people. . . . Cattle raising is a primary factor in the destruction of the world's remaining tropical rain forests," Rifkin writes in the introduction to this book. One underlying theme is that the growing world cattle industry is a major contributor to ecological devastation in many areas of the world, particularly in forest-lands that have been cleared for cattle production. Cattle are produced "at the expense of a burned forest, an eroded rangeland, a barren field, a dried-up river or stream, and the release of millions of tons of carbon dioxide, nitrous oxide, and methane into the skies," noted Rifkin, an environmental activist. His book includes statistics on cattle production—the fourth largest manufacturing industry in the United States, using over 70 percent of the grain produced for cattle feed. The book is also a call for reassessment of the effects of grain-fed cattle production worldwide.

Silcock, Lisa, ed. **The Rainforests: A Celebration.** San Francisco: Chronicle Books, 1990. 223p. ISBN 0-87701-790-5.

This elegant large-format book, with a foreword by Prince Charles, is a pictorial celebration of the world's rainforests compiled by an international team of experts and photographers. Each chapter begins with an overview of a particular web of life in the forest, and a strand from that web is illustrated and described. A strand could be a plant, an insect, a bird, a reptile, or a mammal. The last chapter describes the people who make the forest their home and are being threatened by deforestation. Many of the color photographs show aspects of the forest that have never been photographed before and enhance the viewer's appreciation for the rainforest.

Shoumatoff, Alex. **The World Is Burning: Murder in the Rain Forest.** Boston: Little, Brown, 1990. 377p. ISBN: 0-316-78739-6.

Excerpts of Shoumatoff's analysis of deforestation in the Amazon have been published in *Vanity Fair.* He tells the story of the 1988 murder of Chico Mendes, a grassroots activist and leader of a rubber workers organization in Amazonia, placed against a background of the history and economics of the region as well as the ecology. Shoumatoff presents a firsthand account of some of the destruction and cruelty perpetrated by those who exploit the rainforest.

Skutch, Alexander F. **A Naturalist amid Tropical Splendor.** Iowa City: University of Iowa Press, 1987. 232p. ISBN 0-87745-163-X.

An ornithologist by vocation, Skutch presents his observations of the many birds he has seen in the tropics. He also includes his observations of plants and flowers and the specialized insects found in the rainforest. In his preface, Skutch explains "I have mingled factual accounts with reflections on what I have experienced. . . . I try to interpret what I have seen during more than half a century amid tropical splendor." His book is illustrated with sketches.

Stone, Roger D. **Dreams of Amazonia.** New York: Viking Penguin, 1985. 193p. ISBN 0-670-11533-9.

In this clearly written book, Stone presents his shrewd assessment of how Brazilian and American politics play a major role in the Amazon's environmental balance. He points out that intrusion into the rainforest has been under way for several millennia, but more recent efforts to exploit Amazonia have destroyed what the "developers" most valued. Stone's main emphasis is that the rainforest has greater economic potential if used in its natural state rather than treated as a wilderness to be conquered.

Warburton, Lois. **Rainforests.** San Diego: Lucent Books, 1991. 96p. ISBN 1-56006-150-2.

Written for students in grades 5 through 9, this straightforward book explains the disappearance of tropical rainforests and why the destruction can have dire consequences. Chapters cover familiar topics such as how rainforests supply materials for medicinal drugs, foods, and other products. The book also shows how plants and animals are threatened with extinction and suggests possible solutions to rainforest problems.

Webb, L. N., and Kikkawa, J., eds. **Australian Tropical Rainforest: Science-Values-Meaning.** Melbourne, Australia: CSIRO, 1990. 185p. ISBN 0-643-05055-8.

This is a collection of papers from the fifty-seventh Congress of the Australian New Zealand Association for the Advancement of Science. The contributors are mainly Australian biologists, natural resource managers, and sociologists. Most of the papers are carefully written introductions to rainforest ecology for the nonprofessional reader as well as ecologists and students of environmental politics. The book is organized into four sections: The Depths of Time, Natural Processes, Values and Meaning, and Perspectives for the Future.

Whitmore, T. C. **An Introduction to Tropical Rain Forests.** Oxford and New York: Oxford University Press, 1991. 226p. ISBN 0-19-854276-3 (paper).

Written by a senior researcher at the Oxford Forestry Institute, this book is primarily an introductory text on tropical rainforests for the general college student. It also provides information that can be easily understood by a broad readership. Chapters describe the climate, formations, and growth cycles in the forests; plant life and seasonal rhythms; diversity of rainforest animals; and the interconnections between plants and animals. The text also covers tropical rainforests through time—patterns of distribution of plants and animals—from geological periods millions of years ago to the present. Diagrams, charts, maps, and black-and-white photographs supplement the text. Although many scientific terms are defined within the text, a glossary is a helpful addition as is the "Index to Plants and Forest Products."

————. **Tropical Rain Forests of the Far East.** Oxford and New York: Oxford University Press, 1985. 352p. ISBN 0-19-854136-8.

One of the few textbooks on the tropical rainforests of the Far East, this second edition is a revised version. It is intended for readers who may not have specialized training in biology, so technical language is kept to a minimum. There are five sections: Climate and Seasonality, Growth of Forests, Man and Tropical Rainforests, and two final sections on future trends.

Young, Allen M. **Sarapiqui Chronicle: A Naturalist in Costa Rica.** Washington, DC: Smithsonian, 1991. 384 p. ISBN 1-56098-014-1 (cloth); 1-56098-047-8 (paper).

Young, an entomologist and curator at the Milwaukee Public Museum, provides an informative and readable account of his two decades of experience in the rainforests of Costa Rica. He enthusiastically describes his field studies, exploring and recording the secrets of exotic plants and animals in the rainforests, and shows the interrelated nature of plant and animal life, explaining why forest ecosystems should not be fragmented. He also details the rapid destruction of rainforests in Costa Rica and the loss of plant and animal life that results.

7

Selected Nonprint Resources

THE NONPRINT RESOURCES listed in this chapter include computer databases that can be accessed directly, information services (available for a subscription fee) that provide access to databases, and computer networks through which members can access databases and conferences on a great variety of topics, including those relating to environmental issues such as rainforest destruction and preservation.

In addition, audiovisual materials—computer programs and interactive videodiscs, films, filmstrips, and videocasettes—on rainforests are listed. Addresses and telephone numbers for the suppliers of nonprint resources are provided at the end of this chapter.

Online Computer Databases

Anyone with a personal computer, modem, and communication software can access a database either by dialing a telephone number of a database directly and keying in a password or by subscribing to a computer information service or network such as Dialog Information Services or CompuServe. This sections describes a selection of databases that contain information on rainforests or rainforest-related topics.

AV-Online

The National Information Center for Educational Media compiles this database of information on nonprint media for all levels of educational use, which can be accessed through CompuServe and Dialog. Many audiovisual materials (from audiotapes to videos) on ecology, including rainforest biomes, are indexed and annotated.

EDRI Directory

The EDRI directory is a compilation of significant organizations and foundations that provide grants for environmental purposes. Produced by the nonprofit Environmental Data Research Institute, the electronic database includes profiles of grantmaking institutions and several sample grants. The database can be accessed directly for a fee. A print version of this directory is also available. The institute provides information on either the print or electronic version.

Enviroline

This bibliographic database is available through Dialog Information Services. It covers environmental topics, including rainforest issues, published in 3,500 journals and is updated ten times each year.

Environmental Periodicals Bibliography

The International Academy at Santa Barbara (IASB), California, maintains this database that consists of over 460 environmentally oriented magazines and journals dating back to 1972. For a fee, IASB conducts a search of the Environmental Periodicals Bibliography (EPB) database by subject, author, journal, or geographic descriptor (or any combination of these descriptors) and provides bibliographic information on articles retrieved. The 20-volume EPB is also available on CD-ROM.

Life Sciences Collection

Produced by Cambridge Scientific Abstracts, this database contains bibliographic citations for and abstracts of literature in major areas of biology, biosciences, and biotechnology. The database covers over 5,000 publications, including periodicals, conference reports, and books. Topics such as biodiversity, rainforests, and endangered species can be researched in this collection available through Knowledge Index and Dialog Information Services.

Magazine Index

Produced by Information Access Company (IAC), this database indexes more than 400 popular U.S. and Canadian magazines, including coverage of all magazines indexed in the *Reader's Guide to Periodical Literature.* References to many articles on rainforest topics can be found in this database available through various computer information services.

PAIS International

The Public Affairs Information Service of New York produces this database, which is an international bibliographical index of magazine articles, books, government documents, public and private agency publications, yearbooks, and directories in six languages (English, French, German, Italian, Portuguese, and Spanish). Most of the records include brief abstracts and cover a broad range of social science topics, including government policy and legislation on environmental issues and international trade relations that could have environmental impacts. Access to the database is available through Dialog Information Services, CompuServe, and other vendors.

Computer Networks

Like electronic information services, computer networks usually charge an annual or a one-time fee for access to a network, with additional charges for the time used online. Networks offer access to databases and online conferences with other network users as well as electronic mail (e-mail) services that allow users to create and transfer messages to individuals or groups.

Global Action Information Network

Global Action Information Network (GAIN), as its name implies, is a networking and information service. Its mission is "to provide support to its members so they can take individual as well as collective steps to effect sustainability—in their personal lives, their communities, and the world at large." GAIN electronically publishes environmental issues reports, including several on the world's rainforests, and reviews of proposed legislation and suggestions for actions citizens can take to express their views effectively to elected officials. Access to GAIN is available through EcoNet, one of the networks affiliated with the Institute for Global Communications (see next entry).

Institute for Global Communications

A division of the Tides Foundation, the Institute for Global Communications (IGC) was founded in 1987 to bring together as one organization a peace network (PeaceNet) and EcoNet, an environmental network. IGC now includes ConflictNet and LaborNet as well. Those who subscribe to IGC can access EcoNet and its great variety of conferences (many of them international) that cover environmental issues ranging from bio-diversity to rainforests to toxic waste. Conferences include those spon-sored by the Rainforest Action Network and local Rainforest Action Groups. Several conservation newsletters are online as are reports from the UN Commission on Sustainable Development, environmental law initiatives from around the world, action alerts on environmental con-cerns, and conferences pertaining to indigenous people.

Computer Programs and Videodiscs

Earthquest Explores Ecology
Type: Computer program
Date: 1991
Cost: $69.95
Source: Davidson & Associates

This computer program for grades 5 through 12 features an on-screen South American rainforest to teach climate, food and chemical cycles, biomes, and environmental adaptation. There are a variety of program components including such major areas as the "Rainforest," "Eco-Explorer," and "EcoSimulator." Using the Rainforest menu, students can access screen displays showing forest levels, endangered species, and various flora and fauna. On-screen maps, graphs, a glossary, and a bibli-ography also are available, as are games and animal sounds to enhance and reinforce the program content.

Race To Save the Planet
Type: Interactive series (computer program and videodisc)
Date: 1991
Cost: $395
Source: Scholastic, Inc.

Geared for students in grades 5 through 12, this Nova Interactive Series includes eight program disks (3.5″) and one double-sided videodisc for the Macintosh equipped with 6.0.5 or higher system, 1MB memory, 20MB hard drive, monitor, one floppy disk drive, and videodisc player. The series focuses on both beneficial and detrimental environmental practices with information presented in text, video images (some with sound), and interactive segments. With the use of icons and menus stu-dents make choices in regard to a variety of environmental topics such as habitats/species and case studies of activism such as "Planting Trees in Kenya." A handbook of step-by-step instructions, teacher's guide, and lesson plans are included with the series.

Rain Forest

Type: Computer program and videodisc
Date: 1991
Cost: $285
Source: National Geographic Society

Part of a single-topic video series, this video package targets fifth-through twelfth-grade students. It includes one double-sided videodisc and two program disks (3.5″) for the Macintosh equipped with system 6.0.5 or higher, a minimum of 1MB of memory, hard disk, HyperCard, monitor, mouse, and videodisc player. The program focuses on the rainforests of Costa Rica and is divided into chapters that describe and illustrate rainforest habitats as well as the future of rainforests. Closeup views of various rainforest animals, graphs, charts, and maps are also included.

Films and Videocassettes

Amazonia

Type: VHS videocassette
Length: 70 min.
Date: 1991
Cost: Purchase $95, rental $45; individuals and low-income groups: purchase $39.95, rental $20
Source: The Video Project

This video gives voice to native people of the Amazon forest as well as riverine dwellers, rubber tappers, and small-scale farmers. All of these groups depend for survival upon the forest along the Amazon River. The film brings together first-person accounts of people's attempts to save their homelands, stunning photography of the forest, animation, and original music. It includes a 92-page resource book and is recommended for high school to adult audiences.

Ancient Forests

Type: 16mm color film, VHS videocassette
Length: 25 min.
Date: 1992
Cost: Purchase $390 (film), $110 (video)
Source: National Geographic Society

Suitable for grades 7 through 12, this film takes the viewer through the Tongass National Forest and Prince of Wales Island in Alaska to northern California. The conflict over the logging of the ancient trees is the focus of the film. It presents the view of logging companies, which argue that forest management techniques, including reseeding and promoting natural regrowth, make up for the loss of old-growth trees. But conservationists counter that it takes centuries to replace the giant trees and that continued logging will destroy an extensive ecosystem upon which wildlife and fish depend.

Banking on Disaster

Type: VHS videocassette
Length: 78 min.
Date: 1988
Cost: Purchase $450 ($1/2''$ video), $550 ($3/4''$ video); rental, $50
Source: Bullfrog Films

Filmed over a ten-year period, this award-winning documentary shows the destruction of the Amazon rainforests. It documents the disastrous consequences of cutting and paving a road through the world's largest rainforest. The program presents several points of view, including those of indigenous people and industry and government officials.

Blowpipes and Bulldozers

Type: 16mm color film; VHS videocassette
Length: 60 min.
Date: 1988
Cost: Purchase $850, rental $85 (film); purchase $350 (video)
Source: Bullfrog Films

A highly acclaimed film, this is story of a unique tribe of rainforest people called the Penan, who have lived for over 40,000 years in the jungle of Sarawak, Borneo, part of Malaysia. Bruno Manser from Switzerland spent five years with the Penan, adapting to their lifestyle and helping to bring in an Australian film crew to document the story of the Penan before they are forced out of existence by logging companies supported by the Malaysian government. Manser is now wanted by the government for helping the Penan organize to resist the destruction of their rainforest home and for publicizing the plight of an indigenous people. The film has been called a "must see" for anthropology students and is highly recommended for high schools and public libraries.

Burning Rivers

Type: VHS videocassette
Length: 28 min.
Date: 1992
Cost: Purchase $85, rental $40; individuals and low-income groups: purchase $35, rental $20
Source: The Video Project

The link among the environment, economics, and social problems is the basic theme of this film about the destruction of Guatemala's rich rainforests, the pollution of its rivers, and the health threats to farmworkers from poisonous pesticides. The program shows how serious environmental problems in Guatemala stem from unequal distribution of wealth and land. Wealthy landowners control the most productive land, growing crops for export, while peasants are forced to cut down or burn forests to grow the food they need for survival. Many Guatemalans are forced to live in garbage dumps. In spite of death threats and repression, some Guatemalans have organized to fight injustice. The video comes with a 32-page discussion guide.

The Business of Hunger
Type: 16mm film
Length: 28 min.
Date: 1985
Cost: Purchase $19.95
Source: Maryknoll World Films/Media Relations

This film focuses on peasants who are forced off the land they farm onto rainforest land, which must be cleared. But the cleared forestland is less fertile and must be forsaken for still other areas that will produce food. Eventually, many of the peasants end up in city slums.

Can Tropical Rainforests Be Saved?
Type: VHS videocassette
Length: 120 min.
Date: 1992
Cost: Purchase $19.95
Source: Pacific Arts

Written and produced by Robert Richter, this documentary first appeared on PBS stations and combines investigative reporting with spectacular film of rainforest habitats. This global look at rainforest destruction was filmed in over a dozen rainforest countries as well as in Japan, the world's largest importer of tropical wood. Interviews with business people, government officials, scientists, grassroots organizers, tribal leaders, and many others detail the troubling problems of deforestation worldwide and suggest ways that tropical forests can be saved.

Conservation Biology and Natural Resources Management: Seeking Common Ground and New Directions
Type: VHS videocassette
Length: 40 min.
Date: 1992
Cost: $10 for organizations forming biological diversity chapters

Source: Department of Fishery and Wildlife
            Colorado State University

This video is aimed at undergraduate biology classes, organizations, agencies, and universities interested in conservation biology, and diverse groups concerned with the maintenance of biological diversity. Biologists Thomas Lovejoy and Stanley Temple address topics such as the origin and recent history of conservation biology, the relationship of conservation biology with traditional natural resources disciplines, and the role of students and professionals as advocates.

The Decade of Destruction (classroom version)
Type: 6 VHS videocassettes
Length: 10–19 min. each
Date: 1991
Cost: Purchase 6 programs on one cassette, $350; rental
            $75. Purchase 6 separate cassettes, $450; rental $25 each
Source: Bullfrog Films

These video programs were produced for grades 6–12 by Adrian Cowell with support from the World Wide Fund for Nature. Programs show the sad tale of destruction of the Amazon rainforest during the decade of the 1980s, including devastation of various Indian groups trying to survive the invasion of road builders, colonists, local ranchers, and politicians who want to clear the land and/or tap the vast natural resources of the forests for their own financial gain. Separate programs are "The Rainforest" (10 min.), "The Colonists" (16 min.), "The Development Road" (12 min.), "The Indians" (17 min.), "The Rubber Tappers" (11 min.), and "The Politicians" (19 min.).

The Decade of Destruction (Grades 9–Adult)
Type: 5 VHS videocassettes
Length: 55 min. each
Date: 1991
Cost: Purchase series of 5 cassettes $495, rental $250
Source: Bullfrog Films

Called a "brilliant documentary," the episodes in this series aired on the PBS television program "Frontline." Episodes describe and show real-life stories of people who experienced tragedy and exhibited great courage in their efforts to preserve parts of the great Brazilian rainforest. One of the episodes includes "The Killing of Chico Mendes," the story of the brutal murder of this activist who tried to protect native rubber tappers and their threatened way of life. Other episodes are "The Ashes of the Forest—Parts 1 & 2," "Killing for Land," and "Mountains of Gold."

Earth—The Changing Environment
Type: VHS videocassette
Length: 30 min.
Date: 1987
Cost: Purchase $39.95
Source: PBS Video/Public Broadcasting Service

This film examines the environmental effects of the destruction of the Amazonian rainforest and African farming soils, pointing out the dangers of pursuing economic development without considering environmental factors. Poverty prompts such development.

Earth First: The Struggle for the Australian Rainforest
Type: VHS videocassette
Length: 58 min.
Date: 1990 (revised)
Cost: Purchase $39.95
Source: The Video Project

Commentary from top international scientists is included in this full-color film set in Australia's rainforest. It portrays the struggle to save a 70-square-kilometer stand of forest, which is all that remains of a rainforest that once covered the entire continent.

Ecology: Olympic Rain Forest
Type: 16mm film, VHS videocassette
Length: 20 min.
Date: 1982
Cost: Purchase $350 (film), $275 (video); rental $20 (film)
Source: International Film Bureau, Inc.

The rainforest on the Olympic Peninsula in the state of Washington is the focus of this film that is aimed toward junior high and high school students. It shows how winds from the Pacific Ocean and the topography and location of the area combine to produce an unusually moist and fertile habitat for plants and animals.

Emerald Forest
Type: VHS videocassette
Length: 90 min.
Date: 1985
Cost: Local rates
Source: Local video rental stores

This is a feature film on the effects of a dam project in the Amazon basin and is frequently used in classrooms as well as for home viewing.

Environment under Fire: Ecology and Politics in Central America

Type: VHS videocassette
Length: 30 min.
Date: 1987
Cost: Purchase $35
Source: Environmental Project on Central America
            Earth Island Institute

An excellent educational tool, this film documents environmental destruction in the tropical forests, mountains, hillsides, and agricultural fields of Central America, showing the links among poverty, war, and environmental destruction. The program includes interviews with top Central American and U.S. environmentalists.

Firewood: The Other Energy Crisis

Type: 16mm color film; VHS videocassette
Length: 15 min.
Date: 1984
Cost: Purchase $225 (film), $175 (video)
Source: Dick Young Productions

The link between deforestation and the need for a source of fuel is the main emphasis of this film, which illustrates how people in many parts of the world need firewood for cooking and heating, and are cutting trees to meet these needs.

The Forest through the Trees

Type: VHS videocassette
Length: 58 min.
Date: 1990
Cost: Purchase $85, rental $45; individuals and low-income groups:
            purchase $39.95, rental $25
Source: The Video Project

Produced by former NBC News bureau chief Frank Green, this film focuses not on tropical rainforests but on the last remaining stands of virgin redwoods in the United States and the endangered spotted owl. During the film, loggers, timber company officials, politicians, environmentalists, and local residents present their views and concerns. Besides the competing interests of these various people, the film explains alternatives to current logging practices and policies regarding natural resources in the United States.

Forests Are More Than Trees

Type: Filmstrip/audiotape or slide/audiotape
Length: 20 min.
Date: 1988

Cost: Purchase filmstrip/tape $26.95, slide/tape $29.95
Source: National Wildlife Federation

For an overall look at forests of various types, this audiovisual presentation covers the ecology and threats to forests, such as acid rain and deforestation. The narrative describes the diversity of animals and plants in the forest areas that cover almost one-third of the continental United States.

A Green Earth . . . or a Dry Desert
Type: VHS videocassette
Length: 18 min.
Date: 1990
Cost: UK currency required
Source: Viewtech Film & Video

This video for intermediate through high school students begins with the South American Indian legend that says the tropical rainforest holds up the sky, so if the trees are destroyed, disaster will occur. Based on this theme, the film shows how people around the world depend upon forests not only for a variety of products and biodiversity but also for climate control.

The Kayapo
Type: VHS videocassette
Length: 52 min.
Date: 1987
Cost: Purchase $198 (series $979)
Source: Films Incorporated Video

This is one of several dozen films in a series that records the social structure, beliefs, and practices of societies in remote areas of the world and shows how these societies are now threatened with extinction by the pressures of a technocratic civilization that is expanding rapidly. In this film, photographers and anthropologists take a look at the life of the Kayapo Indians of Brazil's Amazonian rainforest, who have fiercely resisted settlers invading their region. But gold has been discovered on their land, and this isolated tribe must decide what to do with an income of two million dollars per year.

The Kayapo: Out of the Forest
Type: VHS videocassette
Length: 52 min.
Date: 1989
Cost: Purchase $198
Source: Films Incorporated Video

Continuing the story of the Kayapo (see previous entry), this film shows how the Indians have gained international recognition for their bold political resistance and efforts to reassert their traditional cultural identity.

Keepers of the Forest

Type: VHS and Beta videocassette
Length: 28 min.
Date: 1986
Cost: Purchase 3/4" VHS $365, 1/2" Beta $355; rental VHS $65, Beta $55 (includes shipping and handling)
Source: Norman Lippman ("Keepers of the Forest")

Produced by Norman Lippman, this award-winning documentary that has been aired on the Discovery Channel explores the complex causes and far-reaching effects of tropical rainforest destruction in Central America. It focuses on an indigenous group, the Lacandon Maya, in the Lacandon jungle in southern Mexico. The film first looks at the economic pressures that have led to large-scale clearing of the forests and then shows how settlers have used agricultural techniques that are unsuitable for the forest ecosystem and climate conditions. In contrast, the Lacandon Maya have developed a productive and sustainable tropical rainforest management system. But continued forest destruction threatens the Lacandons and their way of life.

Medicine Man

Type: VHS videocassette
Length: 105 min.
Date: 1992
Cost: Local rates
Source: Available at most video rental stores

This popular movie with a rainforest conservation theme is the story of two cancer researchers, Dr. Robert Campbell (Sean Connery) and Dr. Rae Crain (Lorraine Bracco), who meet in the Amazon rainforest while searching for a cancer cure. Campbell is a biochemist and Crain works for a pharmaceutical company. During their quest for medicinal plants the two researchers share adventures and eventually fall in love. Filmed in a Mexican rainforest, much of the action takes place—with the aid of specially built platforms and harness systems—high in the forest canopy.

On the Edge of the Forest

Type: 16mm color film; VHS videocassette
Length: 32 min.
Date: 1977
Cost: Purchase $550 (film), $115 (video); rental $50 (video)
Source: Bullfrog Films

Although produced nearly two decades ago, this program's plea for a commonsense approach to human behavior that will preserve the planet is still relevant today. Economist E. F. Schumacher narrated the film shortly before his death. He walks through a forest in Australia, talking directly to the viewer and describing the efficiency of the balanced forest ecosystem. He points out that people need to learn about the interactions of plants and animals and the need for all life forms to support one another rather than ravage the environment in the name of economics and progress.

Our Threatened Heritage: Endangered Tropical Forests

Type: VHS videocassette
Length: 19 min.
Date: 1988
Cost: Purchase $20
Source: National Wildlife Federation

This film presents some solutions to the problems of tropical deforestation along with a concise examination of the rainforest destruction that threatens the global economy as well as the global ecosystem. Designed for high school students or people in ecology training workshops, the film shows how the death of the forests threatens global weather and valuable biological resources for industry and medicine. The video is in full color and includes vivid photographs of towering trees, fascinating plants, and wildlife in the tropical forest.

Our Vanishing Forests

Type: VHS videocassette
Length: 60 min.
Date: 1991
Cost: $29.25 (for grassroots activists)
Source: Public Interest Video Network (PIVN)

Pulitzer Prize–winning author N. Scott Momaday hosts this documentary that examines U.S. Forest Service practices. Momaday's Kiowa Indian heritage teaches respect and appreciation for the land, and in this film he shows that Americans "must live according to the principle of a land ethic. The alternative is not to live at all." A film for activists protesting the U.S. Forest Service's role in the destruction of forestlands, it shows how massive clear-cutting destroys native National Forests, ancient forests, and unique forest ecosystems. Only stumps remain 20 years after a clear-cut. Using rare historical footage, the film examines 100 years of Forest Service practice to see where the agency went astray from the conservationist principles of its founder, Gifford Pinochet.

Pele's Appeal

Type: VHS videocassette
Length: 30 min.

Date: 1989
Cost: Purchase $35
Source: Pele Defense Fund

Native Hawaiians have been struggling to defend their land and religion from a giant power project on the Big Island of Hawaii. The project also threatens a rainforest. This advocacy film shows how people can help with the preservation project.

The Penan: A Disappearing Civilization
Type: VHS videocassette
Length: 20 min.
Date: 1989
Cost: Purchase $35
Source: Rainforest Action Network

Produced by the Endangered Peoples Project, this film offers a close look at the threatened Penan in Sarawak in northern Borneo. It is the story of a people who live in the oldest rainforest on earth but are facing destruction because of logging operations supported by the Malaysian government.

People and Rain Forests
Type: Slide show (140 slides)
Length: 15 min.
Date: 1991
Cost: Rental $25
Source: Cultural Survival

Designed for all ages, this slide show is an introduction to the indigenous peoples who live in the world's rapidly shrinking rainforest areas. It shows how people live in the forests without destroying them, and includes an easy-to-follow text and maps. The slide story concludes with positive steps being taken in support of the rainforest cultures by groups like Cultural Survival.

Planning for Survival
Type: VHS videocassette
Length: 20 min.
Date: 1992
Cost: UK currency required
Source: Viewtech Film & Video

Produced by the International Centre for Conservation Education, this video explains how people have increased their demands upon global natural resources and are reducing the earth's capacity to support living things, including human life. The film looks at the three main goals of the World Conservation Strategy to maintain the earth's ability to support life, to preserve genetic diversity through the conservation of species, and to ensure the sustainable use of all natural resources.

Preserving the Rainforest

Type: ³/₄″ and ¹/₂″ videocassette
Length: 24 min.
Date: 1991
Cost: Purchase $149, rental $75
Source: Films for the Humanities & Sciences

Similar to many films on rainforest destruction, this one shows the destruction of the Brazilian forest. But in contrast, it also depicts efforts to preserve the forest of Tai on the Ivory Coast. The Tai forest is being used as a natural reserve and as a site for controlled agricultural and industrial activity.

Rain Forests: A Part of Your Life

Type: Enhanced video
Length: 24 min.
Date: 1990
Cost: Purchase $99
Source: SRA School Group

Scientists suggest ways that people and rainforests can coexist in this video, which is a filmstrip that has been enhanced electronically to provide interesting visual effects. The program explains how rainforests absorb carbon dioxide and release the oxygen needed for survival and how forests could help control the greenhouse effect.

Rain Forests: Proving Their Worth

Type: VHS videocassette
Length: 31 min.
Date: 1990
Cost: Purchase $85, rental $25; individuals and low-income groups: purchase $35, rental $20
Source: Cultural Survival

Narrated by Jane Alexander, this film shows the types of products grown or made in the world's rainforests—one way to maintain a sustainable economy. Footage covers various rainforest regions and reveals the problems encountered as producers try to get fair prices for their goods.

RAN Slide Show

Type: Slide show with script
Length: 80 slides
Date: 1992

Cost: Purchase $85, rental $25 (plus $4.50 shipping and handling and $85 deposit, refunded on return)

Source: Rainforest Action Network

This slide show and script describe the ecology of the rainforest and the type of destruction that has taken place. In line with the objectives of RAN, the slide show also explains what people can do to preserve rainforests.

The Temperate Rain Forest
Type: 16mm film, VHS videocassette
Length: 16 min.
Date: 1982
Cost: Purchase $335 (film), $95 (video); rental $30 (film)
Source: Bullfrog Films

This curriculum-oriented film of the rainforest along the U.S. coast in the Pacific Northwest shows the high canopy of tall trees and woodland floor with its tangled undergrowth, home of such animals as salamanders and frogs. There is minimal narration, but information is clear and precise. A good introduction to the temperate rainforest and the impact people have on such an ecosystem, this film is suitable for environmental science classes from junior high school through college.

Tropical Rain Forest
Type: 16mm film, VHS videocassette
Length: 13 min.
Date: 1991
Cost: Purchase $350 (film), $250 (video); rental $75
Source: Cornet/MTI Film & Video

Called "the optimal environment in which life can succeed," the rainforest is the home to an astonishing number of plant and animal species, which are portrayed in this film. The program also explains how heavy rains and brilliant tropical sunshine make the rainforest possible. Through the lush photographs, the viewer explores the multitude of life in the four distinct layers of the forest: the emergent layer at the topmost crown of the rainforest, the canopy layer that forms an umbrella over the forest floor, the understory, and the sparsely populated forest floor.

Unknown Forests
Type: Slide show
Length: 15 min.
Date: 1983
Cost: Purchase $75
Source: World Wildlife Fund Publications

The U.S. Industrial Film Festival honored this 116-color slide presentation with its Creative Excellence award. It portrays some of the vast numbers of plant and animal species in tropical rainforests and shows the unique interconnections between plants and animals in the rainforest ecology.

Vanishing Rainforests: Research To Protect the Tropics
Type: VHS videocassette
Length: 27 min.
Date: n.d.
Cost: Purchase $295, rental $55
Source: CRM Films

Narrated by Lloyd Bridges, this film is an introduction to the global problems related to tropical deforestation and how the expanding indigenous population and land use patterns contribute to deforestation. The film shows how education on land use management and research projects under way may help preserve the forest resources. One of the research projects depicted is a study of the Amazon River to find ways to renew the fertility, fish population, and natural pest control, all of which have been reduced in recent years.

Yanomami: Keepers of the Flame
Type: VHS videocassette
Length: 58 min.
Date: 1992
Cost: Purchase $95, rental $45; individuals and low-income groups: purchase $39.95, rental $25
Source: The Video Project

Television star Michael Dorn hosts this documentary that portrays the journey of a group of journalists, anthropologists, and doctors who traveled to a Yanomami village in the Venezuelan rainforest. Considered the last intact indigenous group in the Americas, the Yanomami's way of life has changed little over the centuries, but their existence is being threatened by gold miners, urban developers, and the diseases brought by people of Western cultures. The film includes a brief history of the indigenous people in the Americas, an in-depth look at the Yanomami from a Western perspective, and a plea to preserve the Yanomami culture.

Addresses for Distributors and Vendors

Bullfrog Films
7745 Mohawk Place
St. Louis, MO 63105
(800) 543-3764

Cambridge Scientific Abstracts
Database Services
7200 Wisconsin Avenue
Bethesda, MD 20814
(800) 843-7751

CompuServe
5000 Arlington Centre
Boulevard
P.O. Box 20212
Columbus, OH 43220
(800) 848-8199

Cornet/MTI Film & Video
108 Wilmot Road
Deerfield, IL 60015
(800) 621-2131

CRM Films
2233 Faraday Avenue
Carlsbad, CA 92008
(800) 421-0833

Cultural Survival
215 First Street
Cambridge, MA 02142
(617) 621-3818

Davidson & Associates
3135 Kishiwa
Torrance, CA 90505
(213) 326-7444 or (800) 545-7677

Department of Fishery and Wildlife
Colorado State University
Fort Collins, CO 80523
(303) 491-6909

Dialog Information Services
3460 Hillview Avenue
P.O. Box 10010
Palo Alto, CA 94303-0993
(415) 858-3785 or
  (800) 334-2564

Dick Young Productions
118 Riverside Drive
New York, NY 10025
(212) 787-8954

Environmental Data Research Institute
797 Elmwood Avenue, Suite #3
Rochester, NY 14620
(716) 473-3090 or
  (800) 724-1857

Environmental Project on Central America
Earth Island Institute
300 Broadway, Suite 28
San Francisco, CA 94133
(415) 788-3666

Films for the Humanities & Sciences
P.O. Box 2053
Princeton, NJ 08543-2053
(800) 257-5126

Films Incorporated Video
5547 N. Ravenswood Avenue
Chicago, IL 60640-1199
(800) 323-4222

Forest History Society
701 Vickers Avenue
Durham, NC 27701
(919) 682-9319

Global Action Information Network
575 Soquel Avenue
Santa Cruz, CA 95062
(408) 457-0130

Information Access Company
362 Lakeside Drive
Foster City, CA 94404
(800) 227-8431 or
    (800) 441-1165

Institute for Global Communications
18 De Boom Street
San Francisco, CA 94107
(415) 442-0220

International Academy at Santa Barbara
800 Garden Street, Suite D
Santa Barbara, CA 93101
(805) 965-5010

International Film Bureau, Inc.
332 South Michigan Avenue
Chicago, IL 60604-4382
(312) 427-4545

Maryknoll World Films/Media Relations
Maryknoll, NY 10545
(914) 941-7590

National Geographic Society
Educational Services
Washington, DC 20036
(800) 368-2728

National Information Center for Educational Media
Access Innovations
P.O. Box 40130
Albuquerque, NM 86196
(505) 265-3591

National Wildlife Federation
1400 Sixteenth Street, N.W.
Washington, DC 20036
(800) 442-7332

Norman Lippman ("Keepers of the Forest")
Documentary Project
7745 Mohawk
St. Louis, MO 63105
(314) 725-3313

Pacific Arts
11858 LaGrange Avenue
Los Angeles, CA 90025
(800) 538-5856

PBS Video/Public Broadcasting Service
1320 Braddock Place
Alexandria, VA 22314-1698
(800) 344-3337

Pele Defense Fund
P.O. Box 404
Volcano, HI 96785
(808) 935-1663

Public Affairs Information Service (PAIS)
521 West 43rd Street
New York, NY 10036-4396
(800) 228-7247

Public Interest Video Network
4704 Overbrook Road
Bethesda, MD 20816

Rainforest Action Network
450 Sansome, Suite 700
San Francisco, CA 94111
(415) 398-4404

Scholastic, Inc.
2931 East McCarty Street
P.O. Box 7502
Jefferson City, MO 65102
(800) 325-6149 or
   (800) 541-5513

SRA School Group
P.O. Box 5380
Chicago, IL 60680-5380
(800) 843-8855

The Video Project
5332 College Avenue, Suite 101
Oakland, CA 94618
(800) 475-2638

Viewtech Film & Video
161 Winchester Road
Brislington, Bristol
England BS4 3NJ
+44-0272-773422
   FAX+44 0272-724292

**World Wildlife Fund
   Publications**
P.O. Box 4866
Hampden Post Office
Baltimore, MD 21211
(410) 516-6951

Glossary

agronomy   The application of plant and soil sciences to crop production.

ancient forest   Often called an old-growth forest—a forest that in general has a large number of old trees that have reached "maturity" and no longer produce new wood each year. Ancient forests are part of many coastal rainforests in the U.S. Pacific Northwest.

atmosphere   The layer of gases that surround the Earth.

biome   An ecological region defined by its climate—temperature and the amount of precipitation—and the plants and animals that have adapted to the particular climatic conditions that occur. There are a variety of biomes such as a desert biome, rainforest biome, and tundra biome.

botanist   A person who specializes in the study of plants.

carbon   A natural element that occurs in all organic compounds and many inorganic compounds.

carbon cycle   Circulation of carbon as green plants absorb carbon dioxide from the air and with energy from the sunlight combine carbon dioxide and water to produce glucose, giving off oxygen as a waste product, which is inhaled by animals and people, who in turn exhale carbon dioxide produced when food is oxidized, or "burned" by their bodies.

carbon dioxide   One of the colorless gasses that makes up the air and is released through the respiration of living organisms.

climatologist   A person who specializes in the study of climate.

community   In ecology, all the plants and animals in a particular environment.

conifer   A tree that bears cones, such as a pine or fir tree.

conservation   The wise use of natural resources to reduce waste and loss.

cultivar   A plant produced by cultivation.

deciduous tree   A tree that sheds or loses its foliage at the end of a season.

desertification   The process by which the amount of available soil water decreases or is eliminated because of such human activities as overpumping of wells or using agricultural methods that deplete soil moisture, thereby drying out the soil so that only desert-type plants are able to grow.

developed countries   Industrialized nations, or those that are primarily industrial and industry provides most of the countries' income.

developing countries   Nations that derive most of their income from agriculture and are developing industries that produce manufactured goods for export.

ecologist   A person who specializes in the study of ecology.

ecology   The scientific study of living things and their relationships with their environment and with each other.

ecosystem   All organisms living in a particular physical environment, which together make up a unit that tends to function as one, with changes in any part bringing about changes in other parts of the system.

ecotourism   Travel that promotes environmental awareness and conservation and provides income for local people who have a vested interest in protecting natural resources.

endangered species   Species threatened with extinction.

entomologist   A person who is an expert in a branch of zoology that deals with insects.

environment   The entire area—including physical, chemical, biological, and other factors—surrounding a particular organism or object.

epiphyte   A plant that uses another plant as a host but does not act as a parasite and receives nutrients from the air or traps food (insects) in its leaves.

erosion   The condition in which natural forces such as rain, wind, and gravity, remove soil from the surface of the earth.

evapotranspiration   The process by which evaporation and transpiration release moisture from plants into the atmosphere.

extinction   In terms of species, the total elimination of all members of a group of plants or animals.

FAO   The Food and Agriculture Organization of the United Nations.

fauna   Animals of a region.

flora   Plants of a region.

fossil fuel   Any fuel derived from decayed organisms, such as coal, gas, or oil.

Gaia hypothesis   A theory that the Earth is a self-regulating organism and creates the conditions under which life exists or thrives.

greenhouse effect   A common term for a theory that a buildup of gases in the atmosphere leads to global warming.

habitat   A type of environment suitable for a particular organism and where that organism is likely to live.

Hodgkin's disease   A progressive and sometimes fatal disease in which the lymph nodes, spleen, and often the kidneys and liver become inflamed and enlarged.

holistic approach   In biology or ecology, an emphasis on an entire organism and the interrelatedness of its parts, as opposed to a focus on individual parts that make up the whole.

indigenous   Native to a particular environment.

mammal   An animal that belongs to a class in which the female have milk-producing glands.

manifesto   A proclamation or public declaration of an individual's or a group's intentions.

NGO   Nongovernmental organization.

nonrenewable natural resource   A natural resource that cannot be replaced in its original state.

nurse log   A decaying log that provides nutrients for seedlings that have sprouted on its surface.

nutrient   A substance an organism needs for normal growth and development.

peasant Someone who is part of an agricultural labor force, usually landless and with only subsistence income.

photosynthesis The process that green plants use to absorb sunlight for energy and produce food from carbon dioxide and water.

reforestation Replanting a cut or burned forest or preparing deforested land for tree planting.

sediment Particles of soil and other solid matter that move from their original site to settle on a land surface or at the bottom of a waterway or body of water.

seed tree A tree left during a forest cutting in order to provide seeds to regenerate the forest.

seedling A young tree, usually less than five years old.

silt Particles of soil classified according to varied standards by their size.

silviculture The care and cultivation of forests.

slash-and-burn agriculture A form of agriculture in which trees are cut and burned to clear land for growing crops.

species A category used in biology to classify organisms that consist of a group of individuals that are capable of interbreeding and producing fertile offspring.

subspecies A classification for plants or animals ranking below species.

sustained yield A harvesting method that reaps nonrenewable resources at a rate that can be maintained long into the future.

taxol A chemical found in the bark of yew trees and used in the treatment of cancer.

technology The tools and techniques used to accomplish a purpose, such as to make products or to apply science.

TFAP Tropical Forestry Action Plan.

threatened species Any species that is likely to become endangered.

transpiration The process of giving off vapor.

UN United Nations.

UNCED United Nations Conference on Environment and Development.

UNEP   United Nations Environment Programme.

USDA   United States Department of Agriculture.

USFS   United States Forest Service.

wilderness area   Lands designated by the U.S. Congress or a federal agency as wilderness and set aside for study.

wildlife refuge   A protected wildlife habitat that may be maintained for such recreational activities as fishing and hunting.

zoology   A branch of biology that deals with the study of animals and animal life.

Index

Kathlyn Gay is the author of more than 60 books, primarily on social and environmental issues, several of which are award-winners. Recent books on environmental topics include *Global Garbage: Exporting Trash and Toxic Waste, Caretakers of the Earth,* and *Cleaning Nature Naturally.* Kathlyn Gay and her husband, Arthur L. Gay, a retired elementary school administrator and teacher, live in Elkhart, Indiana.